LARGE ENERGY STORAGE SYSTEMS HANDBOOK

Mechanical Engineering Series
Frank Kreith, Series Editor

LARGE ENERGY STORAGE SYSTEMS
HANDBOOK

EDITED BY
FRANK S. BARNES
JONAH G. LEVINE

CRC Press
Taylor & Francis Group
Boca Raton London New York

CRC Press is an imprint of the
Taylor & Francis Group, an **informa** business

CRC Press
Taylor & Francis Group
6000 Broken Sound Parkway NW, Suite 300
Boca Raton, FL 33487-2742

First issued in paperback 2017

© 2011 by Taylor & Francis Group, LLC
CRC Press is an imprint of Taylor & Francis Group, an Informa business

No claim to original U.S. Government works

Version Date: 20110701

ISBN 13: 978-1-4200-8600-3 (hbk)
ISBN 13: 978-1-138-07196-4 (pbk)

Library of Congress Cataloging-in-Publication Data

Large energy storage systems handbook / editors, Frank S. Barnes, Jonah Levine.
    p. cm. -- (Mechanical engineering series)
  Includes bibliographical references and index.
  ISBN 978-1-4200-8600-3 (hardcover : alk. paper)
  1. Energy storage. 2. Electric power production. 3. Renewable energy sources. I. Barnes, Frank S., 1932- II. Levine, Jonah G.

TJ165.L37 2011
621.31'26--dc23   2011017383

Visit the Taylor & Francis Web site at
http://www.taylorandfrancis.com

and the CRC Press Web site at
http://www.crcpress.com

Contents

Preface

This book is largely the result of the efforts by a group of students at the University of Colorado Boulder who were interested in making an increasing penetration of renewable energy production useful on the electric power grid. This initial work attempted to include issues of reliability, siting, economics, and efficiency. Over time this work looked at energy storage through the lens of increasing penetration of renewable generation relative to economic and carbon impacts to our power grids. This work began with a focus on a utility scale, and we have added significant contributions from others who have been working in the field in order to cover important topics more completely. Many of the issues that were raised years ago on the economics and the difficulties siting energy storage are still being raised today. We believe that in spite of a number of studies to the contrary, energy storage for the generation of electric power will be important when large amounts of wind, solar, and other renewable energy sources are added to electrical power grids so that these resources can provide a significant fraction of the total electrical energy on the system. In this review we have examined a number of ways that large amounts of energy can be stored and then converted back to electrical energy.

It is our hope that these reviews will be of value to a variety of people including policymakers, developers, and students who need a reference on the issues surrounding energy storage to enable a power grid with high renewable generation penetration.

Frank S. Barnes and Jonah G. Levine

Acknowledgments

We would like to express our appreciation to Dr. Frank Krieth for suggesting this volume and helping us recruit authors for the chapters on thermal and battery storage. We would also like to thank the Colorado Energy Research Institute (CERI), Xcel Corporation, and the Renewable and Sustainable Energy Institute (RASEI) at the University of Colorado for financial support that made much of the work leading to this volume possible.

Frank S. Barnes
Jonah G. Levine

Editors

Frank S. Barnes earned a BS in electrical engineering from Princeton University in 1954, and an MS in 1955 and a PhD in 1958 from Stanford University. He joined the University of Colorado in 1959. He has served as chair of the Department of Electrical and Computer Engineering and acting dean of engineering and was a cofounder of the university's interdisciplinary program in telecommunications.

Dr. Barnes was appointed a distinguished professor in 1997, was elected to the National Academy of Engineering in 2001, and received the Academy's 2004 Gordon Prize for Innovations in Engineering Education. He is a fellow of the Institute of Electrical and Electronics Engineers, the American Association for the Advancement of Science, and ICA. His work on a wide variety of research topics has involved lasers, superconductors, and the effects of electric and magnetic fields on biological activities. Dr. Barnes and his students have been working on energy storage and the integration of wind and solar energy into the commercial grid for several years.

Jonah G. Levine holds a BS in biology (2003) and an MS in telecommunications and utility engineering (2007). He has performed research on energy storage and integration with a focus on renewable energy systems under the direction of Dr. Frank Barnes.

Levine gained working experience with the University of Colorado's Electrical Engineering Department at Boulder and also at the Rocky Mountain Institute; the Turner Endangered Species Fund; and a number of wind, solar, biomass, and energy storage development companies. He is currently employed by Biochar Engineering Corporation and the Center for Energy and Environmental Security to develop biochar production equipment and to form working groups to provide food, energy, and climatic security.

Contributors

Frank S. Barnes
Department of Electrical and
   Computer Engineering
University of Colorado
Boulder, Colorado

Carl Begeal
Department of Mechanical
   Engineering
University of Colorado
Boulder, Colorado

Porter Bennett
Bentek Energy LLC
Evergreen, Colorado

Terese Decker
Department of Mechanical
   Engineering
University of Colorado
Boulder, Colorado

Se-Hee Lee
Department of Mechanical
   Engineering
University of Colorado
Boulder, Colorado

Jonah G. Levine
Biochar Engineering Corporation
and
Center for Energy and
   Environmental Security
Boulder, Colorado

Jozef Lieskovsky
Bentek Energy LLC
Evergreen, Colorado

Gregory G. Martin
National Renewable Energy
   Laboratory
Electricity Resources and Building
   Systems Integration Center
Golden, Colorado

Brannin McBee
Bentek Energy LLC
Evergreen, Colorado

Kent F. Perry
Exploration and Production Center
Gas Technology Institute
Des Plaines, Illinois

Isaac Scott
Department of Mechanical
   Engineering
University of Colorado
Boulder, Colorado

Samir Succar
National Resources Defense Council
New York, New York

1

Applications of Energy Storage to Generation and Absorption of Electrical Power

Jonah G. Levine and Frank S. Barnes

CONTENTS

Introduction

Energy storage currently plays an important role in the electric utility industry. On the current electricity grid, storage capacity has been developed to accumulate energy produced by large, less responsive thermal generation plants and then redispatch it based on peak demand. Storage facilities were largely financed by arbitrage (buying energy at a low price and selling it at a higher price). In addition to the benefits derived by utility companies from arbitrage, energy storage currently contributes to reliability, efficiency, power quality, transmission optimization, and black-start functions. Although different end functions of energy storage affect production, the sole purpose of storage is to increase operational flexibility. The various energy storage technologies allow generation to be followed by distribution on demand within the constraints of storage capacity and the systems in which they function.

The electric energy systems of yesterday largely used energy storage to optimize the dispatch of energy from large thermal generation plants and capitalized on the high value of peak demand. The electrical energy system of tomorrow will use storage as one of many tools for aligning nondispatchable renewable energy generation with load demands. In the future, energy storage will continue to fill its earlier roles and expand to facilitate the technologies of tomorrow. Storage will play an increasingly important part in electric utility operations as they face new challenges arising from the introduction of significant renewable energy sources.

Wind and solar energy are available when weather dictates—not on command. This means that generation does not necessarily correspond to demand. For example, no power from photovoltaic cells is available without sunlight; an end user cannot run an air conditioner on wind energy when the wind is not blowing. Additionally a significant number of solar cells installed at residences, small businesses, and other sites contribute only a fraction of the total power required. As a result, utilities must deal with large numbers of small, widely distributed sources that may or may not be available when peak demand occurs.

The reverse challenge arises from energy generated when load demand decreases. Examples of the variabilities of solar and wind energy are shown in Figures 1.1 and 1.2. The rapid variation in solar power as shown in Figure 1.1 and the lack of availability at certain times cause two types of problems. The

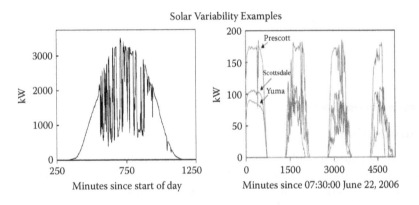

FIGURE 1.1
Variability of solar energy. (Courtesy of AzRise.)

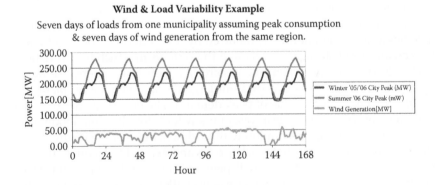

FIGURE 1.2
Electrical load and wind power available for 7-day period in Fort Collins, Colorado.

first is a need to smooth the rapid variations that may occur as small clouds pass overhead (short term variability) and the second is the continuing need for power at nights and on cloudy days (long term variability). These challenges in variability become more significant if the variable (including wind and solar) power of the system exceeds more than 10 to 15% of the total power. When the power fluctuations are less than 10% of the total power of the grid, they may be covered by spinning reserves. However, when a change in load requires additional generation or the supply of renewable energy is more than a system can absorb, changes in system operation are required.

Figure 1.2 clearly shows periods when peak wind power may be available during low loads and little wind energy may be available during high demand. Simply stated, the supply of renewable energy is only partially correlated with the demand for electrical power. During periods of low renewable energy and high demand, the multiple approaches to solving the problem include energy storage, demand response, and bringing gas-fired generators on line. The most common solutions are operating gas-fired generators on line and reducing demand by turning off loads that under agreement may be controlled in this fashion. If a system has hydroelectric power and batteries available, they may also be utilized to meet variations in loads.

During periods when a system has more renewable energy available than it can absorb, the renewable energy may be disconnected from the grid. The details of management of these systems depend on the kinds of power sources available, the loads, and grid structures. Note that power from both coal-fired and nuclear power plants cannot be reduced below a given level without causing serious maintenance, operating, and reliability problems. It may take hours or even days to start coal-fired and nuclear generators after they are shut down. Coal generation may also be limited by allowable ramping rates and tripping off line. Gas-fired generators are susceptible to supply shortfalls and price fluctuations.

Transmission from generation sites to loads may be improved by the capability of dispatching dispatch power from storage on demand. The extent to which energy storage becomes economical to use and its location depend on the system to which it will be integrated. For example, it may be desirable to locate some storage near a wind or solar farm so as to supply peak loads when the wind is not blowing or absorb power when it is not needed. It also may be desirable to locate storage near loads to minimize transmission losses or delay the need to build new transmission lines in areas with growing demands.

As the world embarks on a new era of distributed and at times nondispatchable electrical energy systems, the ability to manage increased levels of variable generation will be critical to each operating region. To manage the increased variability, future electric grids will require flexibility in load and generation management. This variability represents the cost for decreased emissions and increased fossil fuel savings. How utility companies and other energy providers manage the increased variability will depend on the resources available and the variability inherent in their systems. The steps required to achieve effective management for load and generation flexibility include:

- Energy efficiency and demand response
- Spatial and source generation diversity exhibiting complementary profiles
- Ability to bring resources to market via transmission and timely utilization
- Energy storage
- Smart grid electric utility data communications development to integrate above steps

Note that energy storage is only one of these five steps intended to provide flexibility to the energy system. However, energy storage is a key component for ensuring flexibility and reliability with large penetrations of wind and solar energy sources. Pumped hydroelectric systems (PHES), compressed air energy storage (CAES), and other storage systems will facilitate the alignment of renewable generation with loads.

Baseload generation exerts the largest impact on emissions and furnishes the largest portion of energy to a system. If renewable generation is to impact electricity-driven emissions and diversity of supply in a significant way, it must affect baseload generation. When renewable generation comes online coincident with low demand, it will present a challenge because it may not be possible to ramp down the baseload thermal generation systems and require the curtailment of some form of generation. Storage, specifically via PHES and CAES, can address difficulty in ramping rates and help correlate generation and loads.

PHES and CAES take energy from the grid and return it at a later time when it is needed. This raises an important question. What resource is required to power a PHES or CAES facility? Such a resource is most efficient if it is located near a site where a PHES facility pumps power or a CAES compresses air. Thus, when coal is on the margin, it will power a PHES facility; wind on the margin will power a PHES facility. The more renewable energy available, the greater the possibility of having a renewable energy resource on the margin as the prime mover for a PHES facility. The larger the percentage of renewable energy resources, the more flexibility the system will require. At lower penetrations of renewable resources, less storage is needed and this increases the possibility that the storage will be powered by nonrenewable energy resources. At higher penetrations of renewable resources, more storage will be required and this increases the chances that the storage will be powered by renewable resources. Thus, storage is an important issue for developing renewable energy and will reflect decreased emissions by the rest of the system. Renewable energies such as wind and solar generation will benefit from storage as will traditional resources such as coal and natural gas, and other system components such as transmission facilities can benefit from well-placed energy storage. The Electric Power Research Institute (EPRI) lists storage benefits as

- Deferral or avoidance of alternative upgrade or solution net costs that may include components from the transmission, distribution, and generation (TDG) sectors
- Energy costs savings (arbitrage) from the displacement of more expensive peak energy with less expensive off-peak energy
- Transmission peak demand reduction and resulting transmission demand charge reduction for a separate distribution-based utility
- Ancillary services, specifically regulation control and spinning reserve[1]

Storage should be a more significant factor for electric grids as the penetration of renewable generation increases. If the benefits available from spatial and generation diversity and demand response and efficiency are fully utilized, the needs for back-up generation and storage can be minimized. The energy available from intermittent renewable generation combined with the capacity value of storage is a complementary match.

Renewable generation in Colorado, specifically from wind power, is developing rapidly and the development is driving the economy and providing emission-free energy. Facilitating development by adding capacity value would be beneficial. Capacity can be added via:

Energy storage — A storage facility can allow a load to consume excess or low valued energy and deploy high valued energy on demand.

Demand response — Also known as controllable load, this technique may be very effective as shown in the ERCOT report of February 26, 2008.[2]

Overall efficiency — While this factor does not add dispatchable capacity, it is cost effective by reducing baseload.

Increased capacity value or reliability of renewable generation — This can be determined by calculating the probability of generation reliability from spatial and source diversity and optimization planning.[3,4]

Additional natural gas (NG) generation capacity and storage — Although this method is effective, it is burdened with emissions issues. Gas-fired generation can support only the undergeneration of a system and not overgeneration. Moreover NG price fluctuations present economic risks.

Smart grids — Adding capacity by utilizing the steps above can be facilitated via real-time automated communications.

With the introduction of more wind and solar energy to the grid, the addition of a variety of energy storage systems is an option that utilities and developers should consider as a way of increasing reliability and reducing the costs of providing power when needed. The kind of energy storage to be installed depends on both the power required to meet transient needs and the total energy required to match the output of the renewable sources to loads over the times of interest that may be as long as several years to account for weather variations. Fluctuations on the order of minutes (Figure 1.1) might be smoothed by the installation of batteries, supercapacitors, or flywheels. Thermal energy storage may be useful for storing solar energy generated at midday to meet evening demands.

High energy storage applications are currently limited to PHES and CAES. The greenhouse gas output of an electrical system will be a function of the resources running on the system and their respective emissions. A resource on the margin or the next resource in the dispatch order will be the one that powers the energy storage. The higher the penetration of renewable generation in a given system, the more likely the resource on the margin will be carbon-free. Likewise, the greater the penetration of variable renewable generation, the greater the need for firm capacity to respond to loads in the absence of sun and wind.

An addendum to an EnerNex Corporation report on the electrical grid in Colorado shows that at a 10% penetration of renewables without storage (a 324 MW pumped storage facility at Cabin Creek near Georgetown, Colorado), the cost to integrate the renewable generation resources doubled.[5] Although the Cabin Creek facility has proven cost effective and technically capable in wind integration, a modern facility would be more effective by achieving faster response time and greater ability to adjust pumping loads based on the generation available at any given time. A best case scenario combines the traditional pumped storage with advanced variable speed technology. The EnerNex addendum to a report titled "Wind Integration Study for Public Service of Colorado: Detailed Analysis of 20% Wind Penetration"[6] shows a sensitivity analysis of the value of pumped storage relative to the sizing of Cabin Creek. It states:

> In addition to the analysis of wind forecasts, several sensitivity cases were run using the same input data. The Company and TRC felt it was important to include the sensitivities of pumped storage capacity on wind integration costs. This was evaluated by simulating the PSCO system with a varying number of pumped storage units. The existing Cabin Creek units were used as templates: each unit has 163 MW generation capacity and 117 MW pumping capacity. The scenarios modeled were: 0, 1, 2 (current capacity), 3, 4, and 6 units. Generation and pumping capacity as well as the pond size were scaled for each scenario. The integration costs reported here were created using an old (pre-WWSIS), un-smoothed wind forecast (and, therefore, a high geographic diversity of wind farms) and $5/MMBtu gas prices. (Table 1.1)
>
> The results show a decrease in integration cost from about $10/MWh with no pumped storage available to a low of about $3/MWh for 6 units (3 times existing capacity). The implication from this sensitivity analysis is that the ability of pump storage units to pump when there is excess wind and deliver energy to meet the variability needs is of significant value both from an integration cost standpoint and from an overall production cost perspective.

TABLE 1.1

Comparison of Pumped Storage Size Sensitivity Cases

Case Name	Wind Integration Cost ($/MWh) ($5/MMBtu gas)
C0 - No Cabin Creek Units	$10.19
C1 - 1 Cabin Creek Unit	$7.49
C2 - 2 Cabin Creek Units	$5.75
C3 - 3 Cabin Creek Units	$5.34
C4 - 4 Cabin Creek Units	$4.55
C4 - 6 Cabin Creek Units	$2.78

Source: Zavadil, R. M., (2006). "Wind Integration Study for Public Service Company of Colorado". Available online via National Renewable Energy Laboratory, http://www.nrel.gov/wind/systemsintegration/pdfs/colorado_public_service_windintegstudy.pdf (December 5, 2008)

Sullivan, Short, and Blair published a complementary document to the U.S. Department of Energy (DOE) publication titled "Twenty Percent Wind Energy by 2030: Increasing Wind Energy's Contribution to U.S. Electricity Supply."[7] The Sullivan et al. report, "Modeling the Benefits of Storage Technologies to Wind Power,"[8] states, "Given an ideally integrated grid, this [energy storage] capacity would not be necessary [for integration] because the pooling of resources across an electric system eliminates the need to provide costly back-up capacity for individual resources.... It is the net system load that needs to be balanced, not an individual load or generation source in isolation. Attempting to balance an individual load or generation source is a suboptimal solution to the power system operations problem."[9]

The modeling devised to produce the "Twenty Percent by 2030" report lacked the capability to model storage as a component of utility development. Regardless of the capability of the model, the statement suggests that storage for any individual resource on a system scale is not needed; but as the total system variability increases, a resource (storage or other) will be needed to manage that variability, and that leads to an important question. When does a system need a storage resource? Sullivan and colleagues addressed the question and found that at high levels of penetration, storage can lead to more wind power installations and the ability to store electricity adds value to a system as a whole and to wind power in particular.

Four general models were constructed and run to generate comparisons of business-as-usual utility development with and without storage and a 20% wind energy requirement with and without storage. The results showed that business-as-usual utility development with and without the capability to build and use storage allowed an increase in year 2050 wind capacity from 302 GW with no storage to 351 GW with storage. Thus, the additional capacity development of wind without storage is less than the development with storage. These modeled scenarios show the wind and storage development replacing capacity and generation from conventional sources.

In the "Twenty Percent by 2030" cases, the results indicated that with a fixed amount of wind generation, storage can lower electricity prices. Two factors were reported to influence the drop of electricity prices: (1) a reduction in the amount of traditional generation capacity to be built and (2) the ability to store off-peak wind, enabling some wind farms lacking storage to become highly desirable when storage becomes available. The development scenarios forecast by this work showed that storage was brought online only when wind supplied 15% of the nation's energy.

Analyses commissioned by Xcel Energy's Public Service Company of Colorado subsidiary and modeling by DOE's National Renewable Energy Laboratory (NREL) revealed a greater need for energy storage as penetration of wind energy increases. The Energy Storage Research Group at University of Colorado identified two fundamental challenges to renewable wind integration that storage can help address: (1) ramp rate challenges and (2) capacity, further classified as short time scale (under 1 hour) and longer time scale (over 1 hour) challenges. Both types can be addressed with PHES or CAES.

Ramp Rate Challenges

The ramp rate is the short time scale challenge of wind integration. Methods for developing ramp rate curves can be accessed through Levine and Barnes' paper from the 2009 Renewable Power Generation Conference[10] or by contacting the authors directly. Wind generation is not the only variability driver on the grid today; the electric load is variable as well. This implies that (1) utility operators are accustomed to managing variability and (2) load variability plus wind variability represents the load profile that dispatchable generation must meet. In all cases modeled by the authors, the variability of the wind increased the variability of the net load when combined with the load. The curves in Figure 1.2 show the variabilities of 2006 Ft. Collins loads compared with wind energy generation data from typical Colorado wind development locations from 2007 onward, modeled using standard energy generation mathematics.

Table 1.2 quantitatively describes the ramp rates that Xcel/Public Service will need to manage both the current wind system (1 GW) and the proposed system (1.5 GW). The table and figure reflect the same information and indicate that the current system will incur fewer extreme ramping events in a 1-year period than the planned system. To accommodate the increased need for ramping, additional ramping resources should be added to the system's mix (Figure 1.3). Note that the current and proposed systems must ramp both up and down. The ramp rates depend on the generators in the system and can vary from a few minutes for gas-fired turbines and pumped hydroelectric generators to hours or even days to start up coal-fired plants from a cold start.

TABLE 1.2

Histogram of Ramp Events on Current System, Current System plus 480 MW at Peetz Site (Peetz +) and Current System plus 480 MW at Goblers Knob East (Goblers +)

	System of Interest		
	Current	Peetz +	Goblers +
Ramp (MW/hr)	Number of Events		
−1500	0	1	0
−1400	0	1	0
−1300	0	1	0
−1200	0	3	0
−1100	1	2	1
−1000	0	2	0
−900	2	17	5
−800	2	30	14
−700	26	50	42
−600	72	93	89
−500	178	204	195
−400	317	313	337
−300	434	463	420
−200	603	631	655
−100	1010	991	1032
0	1666	1466	1500
100	1632	1517	1537
200	1083	1060	1081
300	769	788	804
400	472	497	481
500	284	321	301
600	146	175	173
700	44	68	56
800	13	33	26
900	5	15	9
1000	1	9	0
1100	0	5	1
1200	0	1	1
1300	0	1	0
1400	0	2	0
1500	0	0	0

Figure 1.4 displays three probability density curves representing the 2006 current, Peetz, and Goblers Knob loads of Xcel/Public Service. The current curve represents an 8760 load curve minus generation of the current wind system modeled at about 1 GW and assessed for ramp rates in 1 hour. Peetz is an 8760 load curve minus generation of the current wind systems plus an

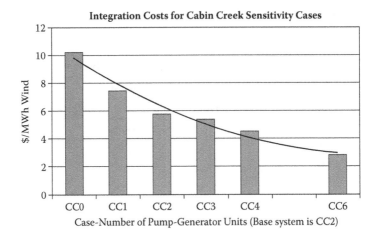

FIGURE 1.3

Example of predicted wind integration cost reduction with addition of storage in Colorado. Pumped storage sensitivity cases at 5 MMBtu gas price. (From Zavadil, R. M., (2006). "Wind Integration Study for Public Service Company of Colorado". Available online via National Renewable Energy Laboratory, http://www.nrel.gov/wind/systemsintegration/pdfs/colorado_public_service_windintegstudy.pdf (December 5, 2008)

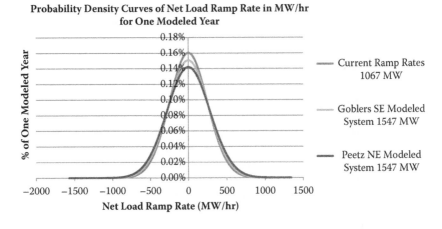

FIGURE 1.4

Probability density curves of net load ramp rate (megawatts per hour) modeled for one year (current, Peetz, and Goblers Knob loads).

additional 480 MW wind generation and assessed for ramp rates in 1 hour. Goblers is an 8760 load curve minus generation at the current wind systems plus an additional 480 MW wind generation and assessed for ramp rates in 1 hour. The three curves indicate that (1) more extreme and frequent ramping occurs with more wind online; (2) the additional capacity in southeast Colorado results in fewer ramping events and less total ramping magnitude than additions in northeast Colorado. Bringing more wind on line, regardless of the location, decreases the probability of consistent hour-to-hour operation and increases the probability of extreme ramping events.

Capacity Challenges

Capacity challenges can be thought of as total magnitude events or longer time scale events. Mitigation of ramp rate challenges through forecasting and other information-based solutions must be addressed when the capacity of a system changes. One example of a capacity event is a general lack of wind generation on a hot summer day when loads are at maximum. Another example of a challenge occurs when the low load hours of spring coincide with high wind generation and create a need to "dump" energy.

Figure 1.5 depicts the load that Xcel/Public Service must meet based on data from 2006. The load is sorted on an annual basis ranging from the highest hour of generation requirement to the lowest. The highest hour requirement is about 8 GW and the lowest is about 3 GW. Three wind generation scenarios are presented. The top line represents the load that must be met without regard to wind generation. Current net load indicated by the middle line indicates the 2006 load curve minus the generation from about 1 GW of wind distributed around Colorado to model the current wind system. The line next to the bottom shows the net load 1500 MW wind in northeast Colorado (current system) with an additional 480 MW of wind generation added to the Peetz geographic area. The net load 1500 MW in southeast Colorado (bottom line) represents the current system plus an additional 480

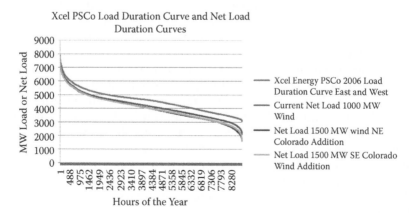

FIGURE 1.5
Xcel Energy/Public Service Company of Colorado load duration and net load duration curves (wind at 1 and 1.5 GW) for 2006.

MW of wind generation at an area known as Goblers Knob East. The charted load duration curves show:

- Wind energy added to Colorado's generation mix can decrease the load that must be met with traditional generation.
- Wind in the modeled scenarios does not decrease the generation needed during peak hours but does decrease the generation needed during minimum (baseload) hours.

The second conclusion is problematic. At the baseload hours (below 3,000 MW), generation must be shut off or turned down, must be sold outside the operational area, or will have to be stored. At baseload level, shutting off or ramping down generation is likely not an option. Selling energy outside the area during baseload hours may mean giving energy away or even suffering a negative energy price. Storing energy would be a solution if storage capacity is available. The current capacity at Cabin Creek is limited to about 1,300 MWh and peak power of 300 MWs in an area where wind generation development is five times greater than the capacity at Cabin Creek. "Common sense and commercial operations finds that the ability to increase Cabin Creek load is invaluable in integrating wind during times when wind generation picks up when load is otherwise low."[11]

It is clear from the foregoing example of integration of wind farms into the grid in Colorado that large utility scale PHES or CAES can facilitate the addition of wind energy to a system and enable the addition to contribute to the baseload. If we assume the extreme case in which there is no wind energy at peak load, then the capital costs for conventional energy sources to meet the peak energy demands are incurred, in addition to capital costs for the wind energy, and the value of the wind farms is limited to reducing fuel costs and greenhouse gas emissions. With geographical diversity of the wind sources, the probability of this occurring decreases. The costs for storage systems must compete with the costs for alternatives such as the addition of gas-fired generators, demand response, and reducing or disconnecting excess wind power.

The addition of a large number of small solar systems dispersed over the grid may help optimize the use of storage systems in a different way. A typical utility network may well evolve from the system shown in Figure 1.6a to the one shown in Figure 1.6b in which multiple types of storage are indicated for use in different parts of the system. At present, batteries would appear to be the most cost effective storage systems to use with small photovoltaic systems in residences. These might be expected to smooth transient demands for a few minutes to a few hours. However, at current prices, it would make little sense to install enough batteries to last for days or longer on systems connected to a grid.

The relatively short lifetimes and high costs of batteries represent significant disadvantages for small stand-alone residential solar systems. The large amount of ongoing research focusing on batteries means their reliability and

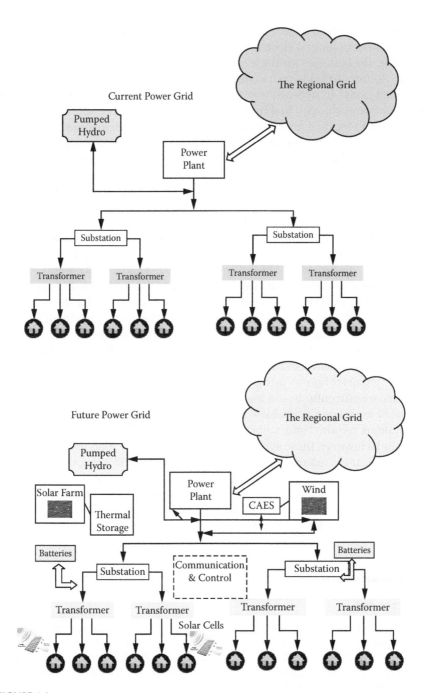

FIGURE 1.6
a: Typical schematic for current power grid. b: Possible schematic for future power grid.

lifetimes are likely to increase and the costs should decrease. Additionally, if a large number of plug-in electric vehicles with proper control systems become available, the batteries for these vehicles may represent a new way to match energy sources to loads on a scale of interest to utilities. The implementation of "smart grids" in two-way communication systems allows management of both loads and sources. Just how much storage may be economical to install and where to locate it are problems that are still under investigation.

At substation level, megawatt level batteries are now available and it is expected that NaS batteries capable of delivering hundreds of megawatts will be available soon.

Batteries in the megawatt range have also been used to delay upgrades of transmission lines. For example, power may be transmitted from a generator to a battery during nights (low load times) and used to supply peak power in afternoons. Additionally batteries of this size are now used to smooth fluctuations in power from wind farms. For large rapid transient needs, fly wheels (not discussed in this volume) are also available. Thermal energy storage seems to fit easily into the development of solar thermal systems and is included in the plans for a number of major projects under development when this book was written. Heat pumps and solar-driven refrigeration systems are now available and undergoing further development.

If a large number of these systems are installed, they will exert an impact on peak air conditioning and some heating loads. For large amounts of energy storage (gigawatt hours), pumped hydroelectric systems and compressed air systems are currently in use and show significant potential as cost-effective means of increasing available storage. Permits for siting pumped hydroelectric systems require time, suitable geological conditions, and large amounts of capital. However, these systems are expected to have long useful lives and can significantly reduce costs of integrating wind and solar energy into a grid. Underground compressed air storage also requires specific geological structures but such sites are expected to be more available than sites for new pumped hydroelectric systems, particularly in the high wind regions of the great plains of the United States.

References

1. Gyuk, I. December 2003. EPRI–DOE Handbook of Energy Storage for Transmission and Distribution Applications. Report 1001834.
2. ERCOT Operations Report on EECP Event of February 26, 2008. www.ercot.com/meetings/ros/keydocs/2008/0313/07.ERCOT_OPERATIONS_REPORT_EECP022608_public.doc
3. Levine, J., and L. Hansen. February 2008. Intermittent renewables in the next generation utility. Rocky Mountain Institute. Renewable Power Generation Conference, Las Vegas, NV.

4. Palmintier, B., L. Hansen, and J. Levine. 2008. Spatial and temporal interactions of solar and wind resources. Next Generation Utility, San Diego, CA.
5. Zavadil, R. May 2006. Wind Integration Study for Xcel Energy/Public Service Company of Colorado. Prepared by EnerNex Corporation, p. 78. http://www.nrel.gov/wind/systemsintegration/pdfs/colorado_public_service_windinteg study.pdf
6. Zavadil, R. December 2008. Wind Integration Study for Xcel Energy/Public Service of Colorado. Addendum: Detailed Analysis of 20% Wind Penetration. Prepared by EnerNex Corporation, p. 45. http://www.nrel.gov/wind/system-sintegration/pdfs/colorado_public_service_windintegstudy.pdf
7. U.S. Department of Energy. July 2008. Twenty percent wind energy by 2030: increasing wind energy's contribution to U.S. Electricity Supply. DOE/GO-102008-2567. http://www1.eere.energy.gov/windandhydro/pdfs/41869.pdf
8. Sullivan, P., W. Short, and N. Blair. June 2008. Modeling the benefits of storage technologies to wind power. American Wind Energy Association Wind Power Conference, Houston, TX. Paper NRE/CP 67043510. http://www.nrel.gov/docs/fy08osti/43510.pdf
9. Ibid., p. 99.
10. Levine, J., and F. Barnes. 2009. An analysis of ramping rates and dispatch timing; matching renewable and traditional energy generation with loads. Rocky Mountain Institute, Renewable Power Generation Conference, Las Vegas, NV.
11. Xcel Energy/Public Service Company of Colorado. 2008 Wind Integration Team Final Report. http://www.xcelenergy.com/SiteCollectionDocuments/docs/CRPWindIntegrationStudyFinalReport.pdf

Suggested Additional Reading

Chen, H. H. 1993. Pumped storage. In *Davis' Handbook of Applied Hydraulics*, 4th Ed. McGraw-Hill, New York, 20.0–20.38.
DeMeo, E. A., G. Jordan, C. Kalich et al. 2007. Accommodating wind's natural behavior: advances in insights and methods for wind plant integration. *IEEE Power and Energy*, November–December, 1–9.
Denholm, P. 2008. The role of energy storage in the modern low carbon grid. DOE–EERE–NREL Energy Analysis Seminar Series, Golden, CO. http://www.nrel.gov/analysis/seminar/docs/2008/ea_seminar_june_12.ppt
Energy Production Research Institute and U.S. Department of Energy. 2003. *Handbook of Energy Storage for Transmission and Distribution Applications*, Washington, Publication 1001834.
Energy Production Research Institute and U.S. Department of Energy. 2003. *Handbook Supplement: Energy Storage for Grid Connected Wind Generation Applications.* Washington, Publication 1008703.
Levine, J. 2007. Pumped hydroelectric energy storage and spatial diversity of wind resources as methods of improving utilization of renewable energy sources. MS Thesis, University of Colorado at Boulder. http://www.colorado.edu/engineering/energystorage/files.html

Parsons, B., M. Milligan, J. C. Smith et al. 2006. Grid impacts of wind power variability: recent assessments from a variety of utilities in the United States. European Wind Energy Conference, Athens, Paper NREL/CP-500-39955. http://www.nrel.gov/docs/fy06osti/39955.pdf

Roza, R. R. 1993. *Compendium of Pumped Storage Plants in the United States: Task Committee on Pumped Storage*. American Society of Civil Engineers, New York.

2

Impacts of Intermittent Generation

Porter Bennett, Jozef Lieskovsky, and Brannin McBee

CONTENTS

Introduction

Intermittent solar and wind energy sources stress the flexibility limits of fossil fuel generation sources to the point where some exhibit severe inefficiencies. Because the utilization of intermittent energy sources is generally mandated by renewable portfolio standards (RPS), system operators must dispatch fossil fuel units to meet total load net of renewable generation, known as net load. Flexible generation sources such as stored energy and natural gas power plants are able to balance the intermittent and volatile generation outputs of wind and solar energy, in contrast to coal facilities that are made more inefficient by irregular operation. When coal facilities are used to balance wind and solar generation, they operate less efficiently. The more wind and solar power used, the more inefficient coal facilities become.

The findings of this chapter are derived from a study conducted by Bentek Energy. The results were published in April 2010.[1] The study utilized hourly generation, fuel consumption, and emissions data from every coal and gas generation unit in the United States with an installed capacity over 25 megawatts (MW). This data, collected and provided publicly by the U.S. Environmental Protection Agency (EPA) under the Clean Air Act and Continuous Emission Monitoring System (CEMS) program, analyzes the impacts of compensating for intermittent generation in the Public Service Company of Colorado (PSCO) and Electric Reliability Council of Texas (ERCOT) service areas.

Of the two intermittent renewable generation sources, solar generation is not as widely used as wind generation. The impacts described in this study focus more on wind than solar operations. Solar facilities operate only when the sun shines and thus avoid the load limitation issues that cause coal plants to cycle.

Wind, Gas, and Coal Integration

Integrating wind generation with generations from other sources presents a number of challenges. The difficulties stem fundamentally from the unpredictability and intermittency of wind. Predictive models show constant improvement but no one can be absolutely certain when wind will blow or for how long it will continue. Historical analyses suggest that wind in the PSCO territory blows most frequently at night. Figure 2.1 compares a wind profile of PSCO's territory published by the National Renewable Energy Laboratory (NREL) to PSCO's average daily load.* Wind generation tends to peak around 4:00 a.m., then declines until about noon before slowly increasing until about 8:00 p.m. The wind peak usually occurs in the early morning hours when system demand (load) is relatively low. PSCO's load, on the other hand, peaks between late afternoon and early evening (2:00 to 9:00 p.m.).

PSCO, like most other utilities, treats wind generation as a "must-take" resource because of the renewable portfolio standard (RPS) mandates. In other words, PSCO will operate its dispatchable resources (coal- and gas-fired plants) in a manner that allows it to take as much generation from wind as possible without allowing generations from its fossil fuel facilities to fall below their design minimum levels.

---

* The wind profile data is from work done in 2008 as part of the Western Wind Integration Study, an ongoing project of the NREL (http://wind.nrel.gov/Web_nrel/). The PSCO load profile represents its average daily loads for 2007 and 2008 based on its Federal Energy Regulatory Commission (FERC) Form 714.

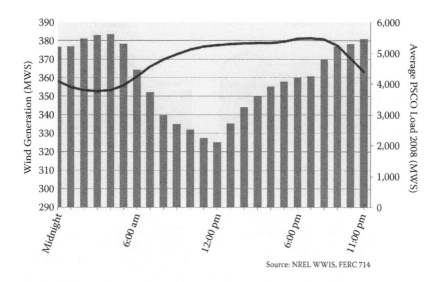

Source: NREL WWIS, FERC 714

FIGURE 2.1
Strongest winds blow between 9:00 p.m. and 5:00 a.m. when power demand is weakest.

When the wind increases, PSCO curtails generation from its dispatchable sources sufficiently to accommodate the wind power. When the wind dies down, generation from the dispatchable sources is brought back online as needed. The process by which generation is ramped up and down at a plant due to wind or any other factor is called *cycling*.

The must-take aspect of wind generation impacts generation stacks differently, depending on the season (Figure 2.2). The solid line indicates the portion of total load that can be met with 1,100 MW of current wind capacity if used at 100% capacity. As shown in the figure, between 8:00 a.m. and 10:00 p.m., coal generation comprises 49% (summer) to 60% (winter) of the generation mix. Accordingly, coal facilities are less likely to be cycled to compensate for wind generation because gas-fired generation (from combined-cycle and combustion turbines) is sufficient to absorb the variability of wind generation. During periods of high load, it is also somewhat easier for PSCO to sell excess power to neighboring utilities to help meet their peak requirements.

After 10:00 p.m., the generation options are different.

Wind resources tend to be strongest at night, when generation from coal comprises approximately 62% of the generation mix and gas-fired generation falls to 20%. If gas-fired generation is insufficient to cycle gas plants safely, coal plants must be cycled instead. Later in the night, when coal-fired generation is the only resource available to absorb wind power, PSCO cycles its coal facilities. As wind energy begins to taper off around 6:00 a.m., the cycled power plants must be ramped up because the daytime load starts building.

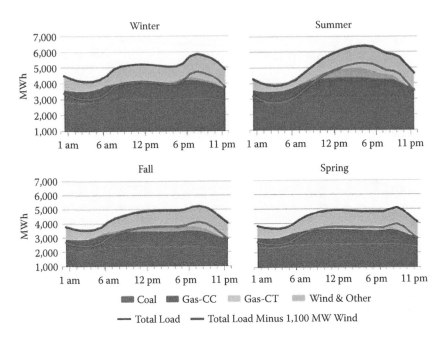

FIGURE 2.2
Impact of wind on power generation stack.

PSCO has another, somewhat restricted, option for offsetting wind generation. It uses its 350 MW of pumped storage hydroelectric power to accommodate wind as much as possible, but when that facility is running at maximum
capacity, it can only operate consecutively for 4 hours.

How frequently wind affects coal- and natural gas-fired generation is difficult to determine because PSCO does not publish hourly wind generation
data.* Nevertheless, PSCO acknowledges wind impacts on both coal and
gas in its addendum to the 2006 Wind Integration Study for Public Service
Company of Colorado.[2,3] In Appendix B of the 2008 addendum, PSCO noted
"a discrepancy between the Cougar modeling and the current experience
when comparing the impacts on coal units. The modeling predicts almost no
impact, but the company [PSCO] is already seeing some cycling that seems
related to wind output."[†]

In other areas of the country, information on wind power is a required
component of power generation reporting. For example, utilities in the
ERCOT area of Texas are required to report their power generation by fuel
type every 15 minutes. Data for 2007, 2008, and 2009 were used to compare coal-plant cycling with wind generation. The analysis identified the

---

* Bentek and the Independent Petroleum Association of the Mountain States (IPAMS) repeatedly tried to obtain 2008 hourly wind generation data from PSCO. All requests were denied
because PSCO contends that the data represent confidential trading information.
† PSCO uses the Cougar model to measure the cost impacts of integration.

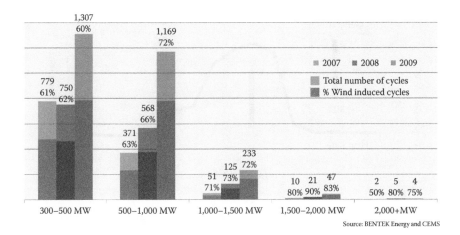

FIGURE 2.3
Distribution of magnitude of ERCOT coal cycling showing hour-by-hour changes.

number of times coal-fired power plants cycled down by 300 to 500 MW, 500 to 1,000 MW, and more than 1,000 MW during the same time periods when wind generation increased by a similar amount. Figure 2.3 shows the results.

In 2009, 1,307 instances in which coal plants were cycled at least 300 MW and 284 examples of cycling more than 1,000 MW from one 15-minute period to the next were reported. Furthermore, the number of instances increased annually since 2007. While Texas has more coal plants and wind farms than Colorado and the Texas wind exhibits somewhat different behavior, this analysis concludes that the two systems are similar enough for a valid comparison. Even in Texas, which has one of the nation's largest gas-fired generation bases, coal plants are frequently cycled. It clearly stands to reason that the same dynamic occurs in Colorado.

Impacts of Cycling

Power plant cycling increases fuel use for every megawatt hour generated. As shown in the first case study discussed in the following section, coal consumption due to cycling exceeded by 22 tons the amount that would have been consumed if the plant had not been cycled (and generation remained stable).

Figure 2.4 depicts operations at PSCO's Cherokee Unit 4* near Denver between 7:00 p.m. and 9:00 a.m. on March 17 and 18, 2008. Total generation

---

* The Cherokee 4 boiler is a 352-MW unit in a 717-MW coal-fired plant in Denver County, CO.

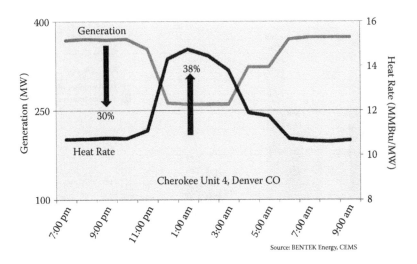

Source: BENTEK Energy, CEMS

FIGURE 2.4
Impact of generation decline on heat rate.

from the plant is shown in the light gray line; the heat rate (MMBtu of fuel per unit of generation) is shown in the dark gray line. Between 9:00 p.m. and 1:00 a.m., generation from Cherokee 4 fell from 370 to 260 MW and then increased to 373 MW by 4:00 a.m. During the period in which generation fell by 30%, heat rate rose by 38%. Heat rates are directly linked to cycling: as coal plant generation falls, the heat rate begins to climb. Initially, the heat rate climbs because generation is choked back and fewer megawatts are produced by the same amount of coal. Later in the cycle, the heat rate climbs further because more coal is burned in order to bring the combustion temperature back up to the designed, steady-state rate; for many hours after cycling, the heat rate is slightly higher than it was at the same generation level before cycling.

As noted earlier, PSCO does not publish hourly wind generation data. However, it publishes hourly generation data for coal plants through its continuing emissions monitoring system (CEMS) reporting. Thus it is possible to examine the behavior of PSCO's coal plants as reflected by their heat rates. Figure 2.5 compares the hourly heat rate versus generation for all coal-fired plants in 2006 to their heat rates in 2008. The average heat rate rose slightly, from 10.45 to 10.57, but overall the total system changed only slightly.

These data, however, mask the impacts on specific facilities. For example, Figure 2.6 compares the hourly heat rates for the Cherokee 4 boiler in 2006 and 2008. Each light gray dot represents the generation and associated heat rate for each hour of operation in 2006 and 2008. The dark gray lines indicate the average heat rates for the boiler during the year. A comparison of the two graphs shows that in 2008 the Cherokee plant was operated in a manner that

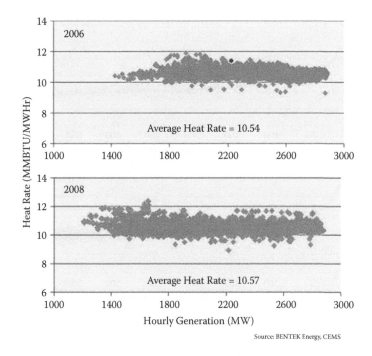

Source: BENTEK Energy, CEMS

FIGURE 2.5
Comparison of heat rate versus generation across all PSCO coal plants for 2006 and 2008.

caused far greater variability in heat rate at different output levels compared to 2006. Why is there a difference?

The only significant change in the operating environment between 2006 and 2008 arose from the addition and use of 775 MW of wind energy. A detailed analysis in a subsequent section discussing two wind events will show concretely how the wind changed operations at Cherokee and other plants. However, these data indicate that cycling coal caused heat rates to become more variable at PSCO's coal plants.

Cycling of coal facilities impacts efficiency and thus affects emissions. To illustrate how cycling a power plant makes its operation less efficient, think about an automobile. When driven at its designed high speed in a high gear, it gets maximum mileage and minimizes emissions. If the driver allows the car to slow without lowering the gear, the car operates less efficiently, decreasing mileage and increasing emissions until it eventually stalls. Conversely, driving at too high a speed for a given gear also makes the car operate less efficiently, resulting in excessive emissions and lower mileage.

A power plant operates in much the same way with only a single gear. Theoretically, coal-fired plants are designed as baseload generators, meaning they are designed to operate at a high utilization rate (typically greater

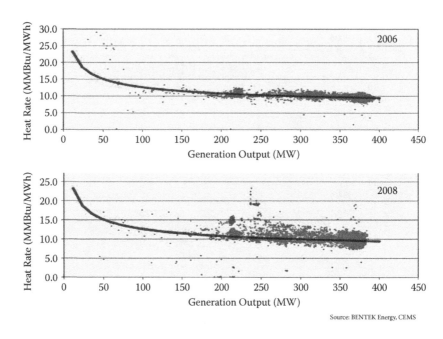

FIGURE 2.6
2006–2008 heat rate changes at Cherokee plant, unit 4.

than 80%) resulting in a flat generation profile. The boilers are "tuned" to combust coal at a specified rate and temperature, and the emissions-control apparatus is synchronized to operate with maximum efficiency at the design rate of the boiler. If the plant must reduce its output, the input rate of the feed coal is cut, thereby allowing the boiler to cool, produce less steam, and thus less power. As long as the boiler is throttled back, it may release fewer emissions simply because it is consuming less coal, but the emission rate (emission per megawatt of output) actually increases because the plant is operating less efficiently.

The emission rate increases further when the temperature of the boiler is increased in order to again increase generation as the wind energy loses strength. More coal must be fed into the boiler to raise the temperature to the design threshold at which it operated before being cut back. In addition, after the boiler is brought back to the desired temperature, the emissions scrubber equipment must be recalibrated and adjusted to achieve optimal control.

The five examples of SO_2 and NO_X impacts from wind events show how emissions rates are impacted by coal plant cycling. Each graphic shows generation during a specific period in the shaded area. Actual SO_2 and NO_X levels are depicted by solid dark and light lines, respectively. The dark and light dotted lines show the average SO_2 and NO_X rates for the month multiplied by hourly generation to derive "normal" emission rates. Days were chosen arbitrarily with the intent of showing the various excess emission patterns that occur after plants are cycled.

In Example 2.1 taken from the CEMS data for the Comanche Unit 1, cycling occurred between 7:00 p.m. on August 17 and 1:00 a.m. on August 18. Generation began to fall around 8:00 p.m.; dropped by 4% between 8:00 and 9:00 p.m.; and dropped an additional 1% between 9:00 and 10:00 p.m. After 10:00 p.m., generation began to build: 4% between 10:00 and 11:00 p.m., and another 3% between 12:00 and 1:00 a.m. About 3 hours later, problems arose with the SO_2 emissions controls that were not restabilized until after midnight. During the night of August 18, total SO_2 output was 16,464 lb higher than if the average SO_2 emission rate had been achieved. NO_X controls appear to have worked well; compared to the average emission rates for the month, the unit generated slightly lower NO_X.

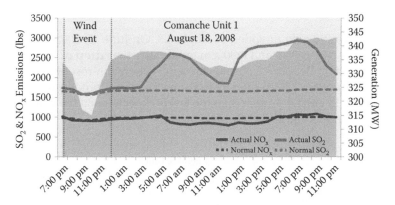

EXAMPLE 2.1

Example 2.2 depicts Cherokee Unit 2 on December 23, 2008, and is more extreme. Between 11:00 p.m. and midnight, generation was reduced by 11%; by 1:00 a.m., generation fell another 30%. It is important to note that this event may well have been triggered by wind due to the sudden steep reduction. Also, these examples show hourly data. In reality, these changes occur from minute to minute and may be even more sudden. As stressful on the equipment as the 24% reduction appears on an hourly basis, the reduction could have been far more problematic if it occurred over a period of a few minutes. After the large decline, production was flat for about 4 hours, rose by 30% between 5:00 and 6:00 a.m., and increased another 13% before 7:00 a.m. Whether this sharp increase occurred smoothly over an hour or happened within a few minutes cannot be determined from the data. In this example, the control equipment worked well: cycling induced 885 lb of additional SO_2 emissions and NO_X

emissions were below average. This example indicates that impacts of cycling were relatively minimal.

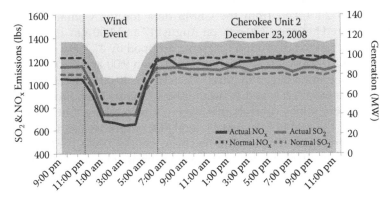

EXAMPLE 2.2

Cherokee Unit 2 also provided Example 2.3. On June 15, 2008, generation fell by 23% between 3:00 and 4:00 a.m., with a further drop of 7% between 4:00 and 5:00 a.m. Between 5:00 and 6:00 a.m., generation rose sharply (14%), followed by another 20% rise before 7:00 a.m. The event produced 3,739 lb of SO_2 and 1,094 lb of NO_x—more than the levels expected had the plant's average emission rate for June 2008 been achieved.

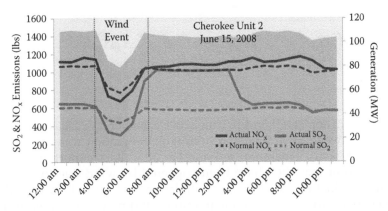

EXAMPLE 2.3

Example 2.4 depicts Cherokee Unit 2 on April 1, 2008. Generation fell by 6% between midnight and 1:00 a.m., and dropped another 22% between 1:00 and 2:00 a.m. Between 2:00 and 3:00 a.m., it rose by 14%, and rose another 20% before 4:00 a.m. This cycling incident generated 1,412 lb of SO_2 and 4,644 lb of NO_x—more than the levels expected had the plant's average emission rate for April 2008 been achieved.

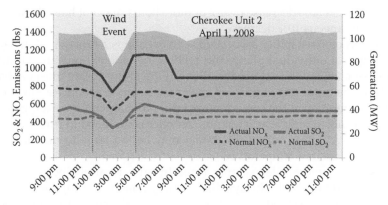

EXAMPLE 2.4

Finally, Example 2.5 depicts generation and emissions at Cherokee Unit 4 for May 2, 2008. Generation fell between 5:00 and 6:00 a.m. by 17%, then fell another 7% before 7:00 a.m. From 7:00 to 8:00 a.m., generation rose by 4%, and then shot up 21% by 9:00 a.m. This event produced 5,877 lb of SO_2 and 1,896 lb of NO_X—more than the levels expected had the plant's average emission rate for the month been achieved.

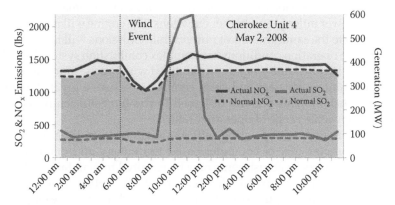

EXAMPLE 2.5

These examples clearly show that cycling causes difficulties for emission control equipment, and that higher-than-normal emission rates continue several hours after a cycling event. It also appears that on occasion emission controls will immediately perform to the extent that emissions initially appear normal, yet problems occur several hours after the event. Cause and effect cannot be determined from this data, but the frequency of these occurrences revealed by the data suggests more than a random relationship. Finally, it is also important to recognize that it is not possible to determine whether the magnitude of the increase or decrease or the suddenness of the event creates the problems. A 30% decrease over 2 or more hours may not have the impact of an instantaneous 10% decrease.

The emissions instability associated with cycling is a function of the ages and designs of individual plants and reflects the inherent operational difficulties associated with coal-fired facilities. If a coal-fired plant must cut back its output, the input rate of the feed coal must be cut to produce a lower rate of steam generation at the correct temperature to maintain low NO_X generation. This is not as simple as it sounds. A boiler is designed to run at certain heat output. At lower output, the boiler may be too large to maintain the output at the desired temperature.

Think of the automobile example again. Imagine a car engine specifically designed to run on flat highways (like a utility boiler). The engine and cooling system were designed to operate at an optimal temperature to achieve the lowest energy consumption and emissions level for the amount of power produced. If you were to drive this car downhill, the engine would generate too much power for the conditions and the engine must be throttled back. With lower power output, the engine would tend to run at a lower temperature because the cooling system was designed to take away far greater amounts of heat than are being generated. Likewise, when the car must run uphill and requires more power, the cooling system may not be capable of evenly cooling the engine. The uneven temperatures within the engine will lead to suboptimal operating conditions. Hot spots in the engine may cause premature ignition, resulting in lower mileage and higher emissions. The engine will require more fuel to generate the same amount of power while emissions will increase.

Varying the operation conditions of a complex combustion system in which a precise and steady flame temperature coupled with precise amounts of fuel and air to maintain efficient and clean combustion poses a great challenge. Boilers are such systems and are designed to run most efficiently within a narrow, steady-state range of operating rates.

The combination of operating efficiently and controlling emissions requires a complex mix of computer-based technology and manual intervention. As many as 50 adjustments may be required to maintain fuel-to-air mixes and lime–slurry mixtures for proper SO_2 absorption in response to changing generation output. Although computerized controls are employed, determining exact adjustments is not always a straightforward process.[4] With changing conditions, the combustion processes are frequently suboptimal and the calculated adjustments may not produce the expected impact on boiler operation. These irregularities cause unstable operation and require manual adjustments—and when manual adjustments must be made, a plant is subject to the greatest risk of instability. Significant emission excesses may result from a suboptimal flame, leading to lower efficiency, partial loss of flame, and in an extreme case, a total plant shutdown.

Another serious consequence of cycling coal plants is plant damage. The financial cost of correcting the damage would include an immediate increase in plant maintenance expense and a reduction of useful plant life—very high costs.* This is especially true for baseload power plants that were not designed to cycle. While it is hard to quantify exactly the costs arising from cycling damage, it is important to include them when projecting wind integration costs. To date, however, most wind integration studies (including those of PSCO) have ignored such costs.[5]

For power plants designed to operate at steady baseload, cycling due to the wind is like driving a car calibrated for the plains of Nebraska in the mountains of Colorado. Such plants will burn more fuel and cause higher emissions. Their operations will cost more over the long run when maintenance and shorter life spans are considered.

PSCO Case Studies

The previous section explained in theory how cycling coal-fired generation plants causes them to operate inefficiently, raising their heat rates and creating a host of other deleterious impacts. This section takes the analysis further by examining two wind events described in detail by PSCO in training materials.

Data and Methodologies

The data employed in these analyses are critical to their credibility. The emission data for CO_2, SO_2, and NO_X derive from the CEMS database maintained by the United States Environmental Protection Agency (EPA). Electric utilities are required to report hourly their total generation, CO_2, SO_2, and NO_X emissions by boiler by plant for all boilers over 25-MW nameplate capacity. Total load is based on data reported by PSCO to the Federal Energy Regulatory Commission (FERC) on Form 714. All control area utilities are required to submit hourly load data.

For any given utility territory, total load data, as reported on Form 714, equals the sum of generation from all plants reported in the CEMS data plus generation from nuclear, wind, hydroelectric, solar, and other non-coal, gas- or oil-generated purchases (spot and contract) from other utilities.

Separating wind and hydroelectric generation on an hourly basis is not possible for PSCO's territory because PSCO is not required to report wind generation beyond monthly and annual levels. As noted in an earlier footnote, PSCO denied requests for 2008 hourly wind generation data and contends that the data represent confidential trading information. Nevertheless, PSCO published hourly data from two dates for studies and training manuals. The

---

* While most plant components are designed to handle cycling, generation changes directly impact water systems, pulverizers, boilers, scrubbers, heat exchangers, and generators. Catastrophic failures resulting from excessive cycling arise commonly from fatigue, corrosion, and cycling-related creep. Such failures may eventually cause plant shutdowns and large capital expenditures for replacement of damaged equipment.

selected days are July 2, 2008, and September 29, 2008.[6] Using the hourly data for days, it is possible to examine in detail how coal, gas, and wind interact and the resulting emissions implications.

July 2, 2008, Wind Event

The first wind event began at 4:15 a.m. and continued through 7:45 a.m. on July 2, 2008. During that period, total wind generation jumped 400% from approximately 200 MW to approximately 800 MW over a 90-minute interval, then dropped back around 200 MW in the next 90 minutes. This event is depicted in the PSCO training manual, shown in Figure 2.7. Coal generation is shown in the light dotted line, wind generation in solid dark line, and gas in light solid line. The jagged black line illustrates the area control error (ACE) used by the National Electric Reliability Council (NERC) to measure system reliability. ACE measures too much or too little power on the system to safely serve total load. In short, it is a measure of reliability. As wind comes online rapidly, ACE spikes upward. Coal generation must be decreased to bring the ACE measure down to the appropriate level.

At the beginning of the event, gas-fired generation accounted for approximately 400 MW or 10% of total load. Coal-fired generation accounted for 2,500 MW or 60% of total load. When the wind commenced, PSCO had to curtail generation at its coal or gas plants to accommodate the incremental wind generation. As shown in Figure 2.7, PSCO chose to curtail generation from coal rather than gas. The motivation for this approach is not clear, but the most likely explanation is that the gas units were operating at near-minimum levels and could not be curtailed further without significant risk to the facilities. To maintain the system margin standards required by NERC, the sudden availability of wind forced PSCO to decrease total coal generation from 2,500 to 1,800 MW, then, back to 2,500 MW in a matter of 180 minutes.

To draw coal-fired generation down, PSCO cycled its Cherokee, Pawnee, and Comanche plants. Figure 2.8 shows the hour-to-hour changes in generation between 4:00 and 5:00 a.m. on July 2, 2008. All PSCO's plants can increase or decrease generation from hour to hour; this hour-to-hour change is known as ramp rate.

As noted earlier, exceeding the designed ramp rate places significant stress on the equipment, renders operations unstable, and potentially shortens equipment life. The hour-to-hour changes shown in Figure 2.8 are compared to the published design ramp rates for PSCO's coal-fired plants as shown in Table 2.1. Cherokee's performance during the incident was within its designed ramp rate; Pawnee operated outside its design rate.

Ramp Rates for Selected PSCO Plants

Operation of the Cherokee coal plant during this wind event illustrates the emission impacts on cycling coal units. The Cherokee plant was chosen due

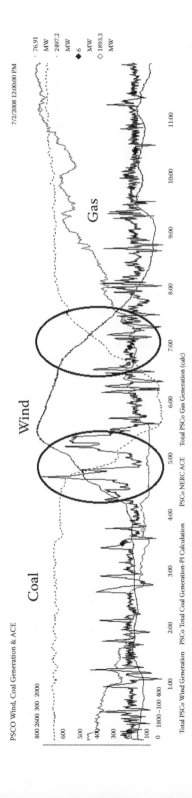

FIGURE 2.7
Wind event impacting PSCO system on July 2, 2008.

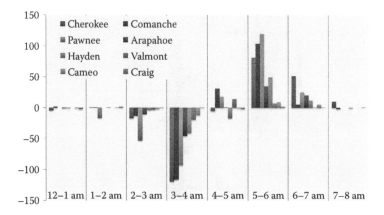

FIGURE 2.8
Hour-to-hour generation changes (MW).

to its proximity to Denver and because it appears to be cycled frequently. The plant contains four coal-fired boilers with summer nameplate capacities of 107, 107 152, and 352 MW. In 2008, the boilers operated at 75, 72, 75, and 83% utilization rates, respectively. Cherokee's hourly generation during this wind event is depicted in Figure 2.9. Between 2:00 and 5:00 a.m., generation fell by 141 MW. Between 5:00 and 7:00 a.m., generation increased until it reached the high for the day of 725 MW at 10:00 a.m. Generation remained essentially flat from about 9:00 a.m. through the balance of the day.

The performance of the coal-fired plant on July 2 contrasts sharply with its performance on July 29 when the system was subjected to less wind and the plant operation was stable. The light gray line in the figure depicts hourly generation on July 29. Although generation declined slightly in the early morning hours on July 29, a rapid decline in generation that occurred on July 2 is clearly not evident. The July 29 curve is shaped very similarly to the curve for the rest of July after the wind event. Total generation on July 29 was 16,603 MWh compared to 16,445 MWh for July 2.

The first step in estimating the emission impact of the July 2 wind event is to calculate the generation as if the event had not happened. A straight line estimates the generation between 3:00 a.m. and 7:00 a.m. if the plant had not been cycled (see Figure 2.10). Generation for the remainder of the day is approximately the same as for July 29 with little wind. Wind generation on the morning of July 2, 2008, caused Cherokee to cycle, reducing generation by 363 MWh.

Calculating Emission Impacts

Three methods were used to estimate the emission impact of the July 2 wind event. The simplest and most common method is to multiply the design emission rates by the generation curve without a wind event (July 29) and by the

TABLE 2.1

Ramp Rates for PSCO's Coal-Fired Plants

Plant	Fuel	Owned or IRP Resource	Capacity (MW)	10-Minute Ramp Rate	
				(MW)	% Cap
Arapahoe 3	Coal	Owned	45	6	13
Arapahoe 4	Coal	Owned	111	5	5
Cabin Creek A	HE	Owned	162	95	59
Cabin Creek B	HE	Owned	162	150	93
Cherokee 1	Coal	Owned	107	6	6
Cherokee 2	Coal	Owned	106	6	6
Cherokee 3	Coal	Owned	152	22	14
Cherokee 4	Coal	Owned	352	20	6
Comanche 1	Coal	Owned	325	22	7
Comanche 2	Coal	Owned	335	22	7
Fort. St. Vrain	NG	Owned	690	75	11
Pawnee	Coal	Owned	505	16	3
Valmont 5	Coal	Owned	186	14	8
Valmont 6	Coal	Owned	43	43	100
Arapahoe 5, 6, and 7	NG	IRP	122	20	16
Blue Spruce	NG	IRP	271	81	30
Brush 1 and 3	NG	IRP	76	18	24
Brush 2	NG	IRP	68	19	28
Brush 4	NG	IRP	135	44	33
Fountain Valley	NG	IRP	238	34	14
Manchief	NG	IRP	261	97	37
Rocky Mountain Energy	NG	IRP	587	103	18
Spindle Hill	NG	IRP	269	119	44
Thermo Fort Lupton	NG	IRP	279	147	53
Tristate Brighton	NG	IRP	132	55	42
Tristate Limon	NG	IRP	63	27	43
Valmont 7 and 8	NG	IRP	79	38	48

HE = hydroelectric. NG = natural gas.

generation curve with the event (July 2), then compare the results over the duration of the event (Method A). Table 2.2 summarizes the calculations. The measured emission rates for July 29 are presented in the first row. The second row indicates total emissions for the no-wind scenario; the third row shows total emissions associated with July 2 generation. Analyzing the emission impacts in this manner results in the estimate that the wind event reduced SO_2 by 730 lb, NO_x by 1,386 lb, and CO_2 by 392 tons (bottom row). The limitation of Method A is that it replaces the actual emissions that occurred on July 2 with estimated emissions from a stable day; they are lower because of the inefficiency of the boiler by cycling as described above.

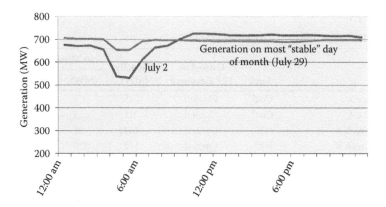

FIGURE 2.9
Actual and projected generation at Cherokee plant.

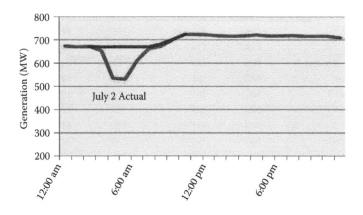

FIGURE 2.10
Actual and projected generation at Cherokee plant on July 2, 2008.

Method B corrects the calculation by substituting the actual emissions on July 2 for the estimated emissions on that date. The emission rates were actually much higher than the "stable day" rates of by Method A and reflect the impact of cycling. Table 2.3 compares the timeframes using the emission rates reported in the CEMS data for the July 2 wind event. Using the actual emissions data yields the result that cycling Cherokee produced 6,348 lb more SO_2, 10,826 lb more NO_X, and 246 fewer tons of CO_2. Method B's limitation is that it focuses only on emissions associated with a specific event, in this case the activities from 3:00 and 7:00 a.m. However, the sudden decrease of generation followed by an increase at the Cherokee plant caused emissions variability that extended well beyond 7:00 a.m. when the plant returned to its pre-cycle generation level. Table 2.4 depicts the additional emission impacts because it includes generation and emission data for all of July 2.

TABLE 2.2

Estimated Emission Savings from Wind on July 2, 2008 (Method A)

	SO$_2$ (lb)	NO$_X$ (lb)	CO$_2$ (ton)
Estimated stable day (July 29) emission rates (per MWh)	2.01	3.82	1.08
Estimated stable emission rates, no wind generated (3:00–7:00 a.m.); total generation = 3,360 MWh	6,754	12,829	3,628
Stable rates, actual generation (3:00–7:00 a.m.); total generation = 2,997 MWh	6,025	11,443	3,236
Saved [additional] emissions	730	1,386	392

TABLE 2.3

Estimated Emission Savings from Wind on July 2, 2008 (Method B)

	SO$_2$ (lb)	NO$_X$ (lb)	CO$_2$ (ton)
Estimated stable day (July 29) emission rates (per MWh)	2.01	3.82	1.08
Actual July 2 emission rates (per MWh)	4.37	7.89	1.13
Stable emission rates, no wind generated (3:00–7:00 a.m.); total generation = 3,360 MWh	6,754	12,829	3,628
Actual July 2 emissions (3:00–7:00 a.m.); total generation = 2,997 MWh	13,103	23,655	3,383
Saved [additional] emissions	[6,348]	[10,826]	246

TABLE 2.4

Estimated Emission Savings from Wind on July 2, 2008 (Method C)

	SO$_2$ (lb)	NO$_X$ (lb)	CO$_2$ (ton)
Estimated stable day (July 29) emission rates (per MWh)	2.01	3.82	1.08
Actual July 2 emission rates (per MWh)	4.37	7.89	1.13
Estimated stable emissions, no wind generated (3:00–7:00 a.m.); total generation= 3,360 MWh	33,787	64,175	18,151
Actual July 2 emissions (3:00–7:00 a.m.); total generation = 2,997 MWh	71,897	129,799	18,561
Saved [additional] emissions	[38,109]	[65,624]	[410]

Method C (Table 2.4) provides the most accurate analysis because it captures the total impact of cycling the plant. The net result is that cycling Cherokee on July 2 resulted in greater emissions, even netting the emission avoided by using wind. Table 2.5 summarizes the results of the three calculation methods. If wind generation had not caused PSCO to cycle Cherokee on July 2, 38,110 lb of SO$_2$ or 53% of the day's total SO$_2$ emissions, 65,624 lb or 51% of

TABLE 2.5

Summary of Calculations of Estimated and Actual Emissions of SO_2, NO_x, and CO_2 via Methods A, B, and C

	SO_2 (lb)	NO_x (lb)	CO_2 (ton)
Method A			
September 22 emission rates (per MWhr generated)	0.0305	0.0320	0.0110
Actual stable emissions generated 8 p.m.–3 a.m.	48,370	50,778	17,457
Estimated stable emissions generated 8 p.m.–3 a.m.	41,900	43,986	15,122
Saved [additional] emissions	6,470	6,792	2,335
Method B			
September 22 emission rates (per MWhr generated)	0.0350	0.0320	0.0110
September 28 emission rates (per MWhr generated)	0.0345	0.0361	0.0112
Actual emissions generated 8 p.m.–3 a.m.	48,370	50,778	17,457
Stable day emissions (no wind) generated 8 p.m.–3 a.m.	47,430	49,580	15,356
Saved [additional] emissions	940	1,198	2,101
Method C			
September 22 emission rates (per MWhr generated)	0.0350	0.0320	0.0110
September 28 emission rates (per MWhr generated)	0.0345	0.0361	0.0112
Actual emissions generated 8 a.m.–3p.m.	160,646	167,926	52,010
Stable day emissions (no wind) generated 8 p.m.–3 a.m.	131,823	150,909	53.969
Saved [additional] emissions	[28,823]	[17,017]	1.686

NO_x, and 410 tons or 2.2% of CO_2 would have been avoided. The use of wind generation in a manner that forced PSCO to cycle Cherokee added significant emissions from the Cherokee plant on July 2, 2008. Additionally, assuming that the same quality of coal was used throughout the event, cycling the plant also required PSCO to burn approximately 22 tons more coal than it would have burned if the plant had not been cycled.

Figure 2.11 also shows how important the definition of event duration is to the estimated impact. If the narrow 3:00 to 7:00 a.m. definition is used, the impact of cycling is considerably smaller. However, this definition does not consider the longer term difficulties of recalibrating the emission controls after a significant cycling event which, as we have seen, can result in increased emissions over several hours. Clearly the longer term perspective involves the most appropriate means to measure these impacts.

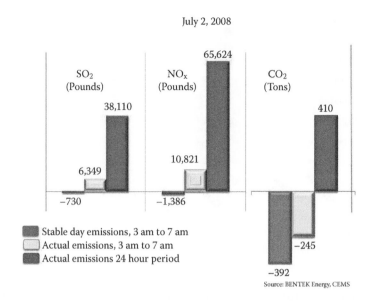

FIGURE 2.11
Incremental emissions resulting from cycling of Cherokee plant on July 2, 2008.

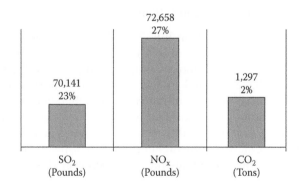

FIGURE 2.12
Incremental emissions impacts of coal plant cycling on all PSCO plants on July 2, 2008.

The same analysis was used to estimate the emissions implications of the July 2 event for all the coal-fired plants in PSCO's resource base. The results are summarized in Figure 2.12. Using the 24-hour event definition (Method C) across the system, the July 2 wind event caused 70,141 pounds of SO_2 (23% of the total PSCO coal emissions), 72,658 pounds of NO_x (27%), and 1,297 more tons of CO_2 (2%) to be emitted than if the event had not caused the plants to be cycled.

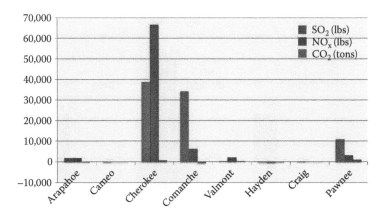

FIGURE 2.13
Incremental emissions by plant on July 2, 2008.

As shown in Figure 2.13, most of the additional emissions came from three plants, Cherokee, Comanche, and Pawnee. All of these plants are located near Denver and thus directly impact emissions levels along the Front Range.

Conclusions Related to July 2, 2008, Wind Event

System-wide, wind generation on July 2 produced 70,141 lb of SO_2 (23% of total) and 72, 658 lb of NO_x (27% of total). Wind generation saved 1,249 tons of CO_2 (2% of total CO_2 emissions). Compensating for wind generation on July 2 appears to have resulted in inefficient and abnormal operation at PSCO's coal plants that resulted in increased total SO_2 and NO_x emissions. By netting out the emissions associated with coal-fired generation that were avoided by using wind, the result is that due to wind generation, SO_2 and NO_X emissions were significantly higher (23 and 27%, respectively) than they would have been if the coal plants had not been cycled to compensate for wind generation.

September 28–29, 2009, Wind Event

The second wind event began during the night of September 28–29, 2008, as depicted in Figure 2.14 from a PSCO training manual. As total load decreased during the night, PSCO reduced generation at coal and gas units to allow wind to continue to generate. When the wind event commenced, PSCO was generating approximately 2,000 MW from coal and 1,500 MW from natural gas. Beginning at 10:00 p.m. on September 28 and continuing until 2:00 a.m. the following morning, coal generation was ramped down by approximately 25% to 1,487 MW until wind generation dropped to approximately 50 MW

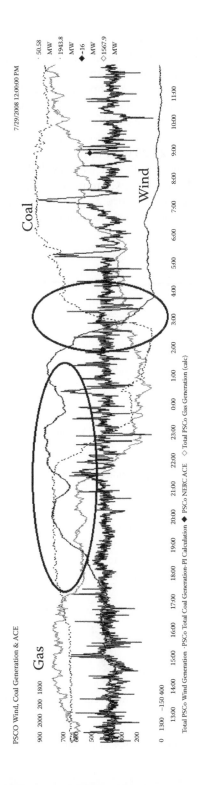

FIGURE 2.14
September 28–29, 2008, wind event impacting PSCO plants.

between 2:00 and 4:00 a.m. In response, coal was ramped up from approximately 1,500 to 1,900 MW in 60 minutes beginning at 3:00 a.m.

Generation from all PSCO coal plants on September 28–29, 2008, contrasts to generation a few days earlier (September 22–23). Figure 2.15 details the hourly generation for both sets of days. Wind generation availability on September 28–29 resulted in a significant reduction in coal-fired generation. As was done for the July 2 case study, the emission rates associated with generation from September 22–23 were applied to the September 28–29 event.

Figure 2.16 shows the plants that were cycled to accommodate wind on September 28–29. The Pawnee, Comanche, and Cherokee coal units were cycled to balance the load. Figure 2.17 shows the coal generation avoided during the wind event, aggregated to include all coal-fired plants. The event

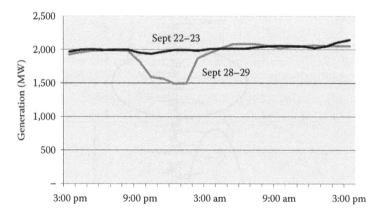

FIGURE 2.15
Comparison of PSCO coal plant generation on September 28–29 and September 22–23, 2008.

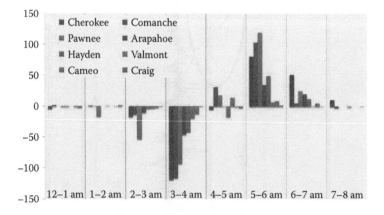

FIGURE 2.16
Hourly generation changes on September 28–29, 2008.

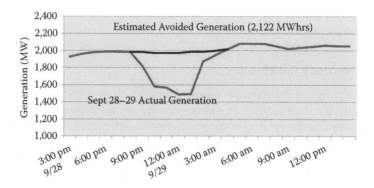

FIGURE 2.17
Estimated avoided generation from wind event on September 28–29, 2008.

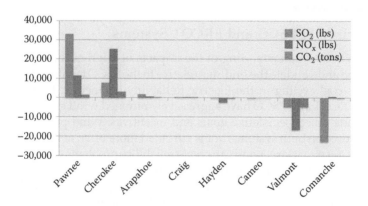

FIGURE 2.18
Distribution of extra emissions by PSCO plant on September 28–29, 2008.

is estimated to have avoided approximately 2,122 MWh of coal-fired generation between 8:00 p.m. and 4:00 a.m.

The estimated extra emissions generated are shown in Table 2.5 using the same three calculation methods described earlier. As with the July 2 event, the calculation method drives the results. If the additional emissions that occurred September 29 (after the wind fell off and coal generation resumed) are included, this wind event resulted in 28,823 lb of SO_2 and 17,017 lb of NO_x (18% of total SO_2 and 10% of total NO_x generated that day) more than would have been emitted had coal not been cycled. On the other hand, using wind to the degree it was used on September 29 allowed PSCO to avoid generating 1,686 tons of CO_2 (3.2% of total).

Figure 2.18 shows the distribution of the emissions associated with the Method C calculation. Virtually all the extra SO_2 and NO_x emissions were created at the Pawnee and Cherokee plants. The Arapahoe, Hayden, and Comanche plants showed small NO_X savings.

PSCO Case Study Conclusions

The case studies in this section conclude that cycling coal-fired facilities to compensate for intermittent must-take energy sources results in inefficient operation during the cycling event and for hours afterward. This inefficiency results in severe degradation of emission savings at best and net additional emissions in many cases. Variable generation sources such as stored energy and natural gas facilities are necessary on systems that utilize intermittent energy sources to fully realize emission savings.

Comparison of PSCO and ERCOT Systems

To gain a better understanding of the impacts of wind events on coal-fired generation and validate the findings relative to the PSCO territory, this section examines coal cycling in the Electric Reliability Council of Texas (ERCOT) system. ERCOT and PSCO have aggressively pursued wind generation in the past decade due to legislative goals and incentives. Wind power is a must-take resource on both systems, but is curtailed more often at ERCOT because resources are much larger and can create reliability problems when the system is fully generating. Finally, both systems are dispatched by central operators that attempt to utilize as much wind generation as possible without disrupting reliability standards. More important than these similarities, however, are the distinctions. ERCOT has far larger gas-fired generation capacity and requires publishing of detailed wind generation data that, when combined with CEMS data, enables precise definition of wind events, thus facilitating a better understanding of the emission implications of wind use.

Wind, Coal, and Natural Gas Interactions in ERCOT System

This section examines the interactions of wind, coal, and natural gas in the ERCOT region of Texas as a means of further validating the results found in the PSCO territory. It will demonstrate that while ERCOT's scale of wind, gas and coal operations is larger than in PSCO's territory, the result is the same. Since the wind blows at night when gas generation is relatively low as a percent of total generation, coal plants are cycled, producing more SO_2, NO_X, and CO_2 than would have been the case had those coal plants not been cycled.

ERCOT publishes wind, coal, nuclear, natural gas, and hydroelectric generation data on a 15-minute basis. In addition, hourly generation and emissions
data are available through the CEMS system. Both the ERCOT 15-minute
data and the CEMS 60-minute data were utilized to determine the emission
implications of cycling units due to wind generation.

The same methodology was used for calculating emission implications of
wind in ERCOT as was used in the PSCO analysis with one exception. Due
to the availability of the 15-minute generation data, wind event details can be
calculated more precisely. For the ERCOT analysis, a wind event was defined
as an instance during which a 10% or greater dip in coal generation coin-
cided with an increase in wind energy generation. The frequency of cycling
events in ERCOT is captured within this section along with a case study
covering a 1-day period.

Frequency of Coal and Gas Cycling

Coal plants in the ERCOT system are cycled based on wind generation. The
8-day example shown in Figure 2.19 illustrates the mechanism. Every day,
as wind increases between 9:00 p.m. and 5:00 a.m., coal generation dips. On
some days, such as November 9 and 10, coal generation drops significantly,
but even on days of limited wind such as November 8, wind appears to push
a small amount of coal generation offline.

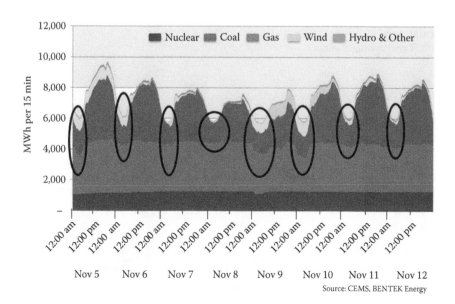

Source: CEMS, BENTEK Energy

FIGURE 2.19
Cycling of coal plants as wind generation increases, November 5–12, 2008.

Figure 2.20 shows the impact of wind on coal cycling. The solid bars indicate the number of wind-induced cycle events. The shaded portion represents cycling events not related to wind. The categories capture the sizes of the events. For example, the first category (300–500 MW) indicates that the number of times total coal-fired generation increased from 300 to 500 MW from hour to hour.

This data indicates that most coal cycling in Texas is due to wind generation and that the number of wind-induced cycling instances is increasing rapidly. Figure 2.21 compares wind-induced coal-cycling events from Figure 2.20 to the total wind generation for each year. In 2008, wind generation grew by

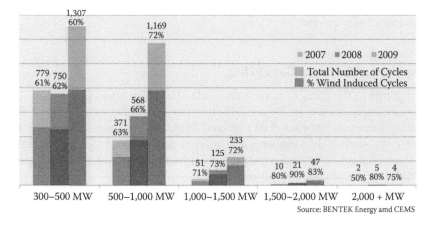

Source: BENTEK Energy amd CEMS

FIGURE 2.20
ERCOT coal cycling events.

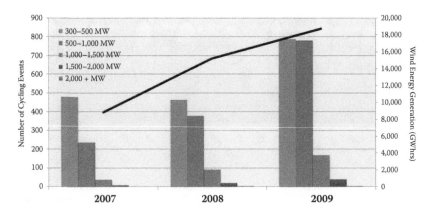

FIGURE 2.21
ERCOT wind-induced coal cycling and wind generation.

73% over 2007 and increased another 23% in 2009 over 2008. The incremental growth in 2009 appears to have had a more profound impact on the incidence of cycling than did the larger growth in 2008. This suggests that the impact of wind is cumulative: the more wind that comes on the system, without corresponding additions of other generation forms, the more wind-induced coal cycling results.

Emission Impacts: J.T. Deeley Plant Case Study

Data for November 8 and 9, 2008, show contrasting generation results. Figure 2.22 illustrates the generation mix for both days. Little wind generation was present on the morning of November 8. Wind accounted for 2% of total generation that day. As a result, coal-fired generation produced power consistently throughout the morning until late evening. About 8:00 p.m. on November 8, wind generation began coming online and grew until it peaked about 7:00 a.m. on November 9. Wind generation was strong throughout November 9 and accounted for 12% of total generation. Coal units were cycled throughout that day to accommodate wind generation.

One coal-fired plant was chosen to illustrate the impact of coal cycling. The J.T. Deeley plant was one of the plants that accommodated the wind on November 8–9. Figure 2.23 details hourly generation and emissions. The graphic shows a sharp drop in generation, beginning about 9:00 p.m. SO_2 initially followed suit and fell until generation began to rise about 4:00 a.m. on November 9. After that, SO_2 rose with increased generation and did not flatten out when generation reached its peak around 7:00 a.m. For

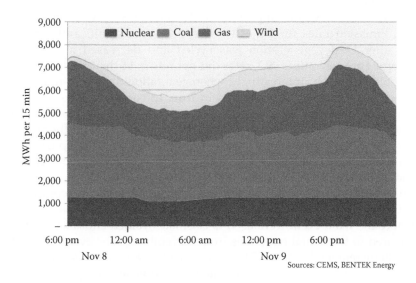

FIGURE 2.22
ERCOT generation mix on November 8–9, 2008.

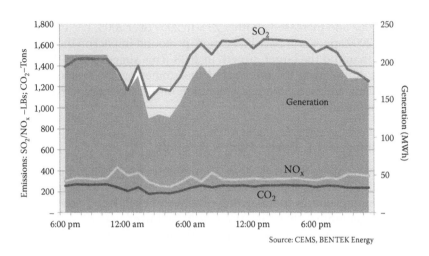

Source: CEMS, BENTEK Energy

FIGURE 2.23
J.T. Deeley plant generation and emissions on November 8–9, 2008.

the remainder of the day, generation held between 199 and 178 MWh—10 MWh below the pre-event generation level—yet SO_2 emissions exceeded pre-event levels by an average of 161 lb until 9:00 p.m. when it finally fell back as generation again declined. NO_X and CO_2 both rose slightly as coal generation fell, but, as the generation came back online, emissions quickly returned to and held at their pre-event levels. The behavior depicted in Figure 2.23 suggests that the emission rates did not fall proportionately to generation.

Figure 2.24 shows the impact of the November 8–9 event on emission rates. Emission rates for SO_2, CO_2, and NO_X rose significantly immediately after Deeley generation was cycled and decreased as generation was brought back online. SO_2 rates did not return to their pre-event levels until late in the day. Interestingly, when generation dropped around 10:00 p.m. on November 9, NO_X rates again rose. In comparison to November 8, emission rates on November 9 are significantly higher. If generation at Deeley remained constant instead of variable on November 9, the emission rates would have been similar to those of November 8. The top left line in Figure 2.25 depicts the 247 MW of avoided generation due to cycling for wind on November 9.

To calculate emissions associated with the event, Method C (discussed in the PSCO section) was employed. The stable day rates evidenced on November 8 before the wind event were used to calculate avoided emissions and then compared to the actual emissions from November 9. The event resulted in 2,506 lb incremental SO_2 and 717 lb incremental NO_X and saved 120 tons of CO_2. Cycling J.T. Deeley to compensate for wind generation produced more SO_2 and NO_X emissions than would have been generated if the plant generated the same amount of power at a flat level. Due to cycling, J.T. Deeley

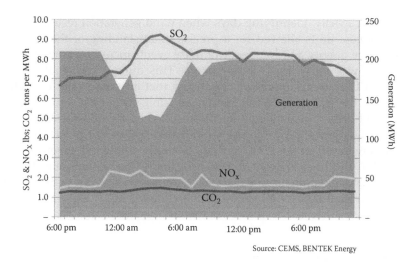

FIGURE 2.24
J.T. Deeley plant generation and emission rates, November 8–9, 2008.

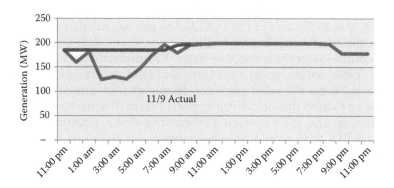

FIGURE 2.25
J.T. Deeley plant generation, November 9, 2008.

emitted 8% more SO_2 and 10% more NO_X while saving 2% of CO_2 emissions. This case study indicates that, like the PSCO examples, coal plants in Texas operate at the highest efficiency during steady-state operations at the levels for which they are designed. Operating these facilities irregularly or at non-design levels leads to inefficient operation and higher emission levels.

Conclusions Related to ERCOT Operations

The ERCOT system was studied due to the availability of wind data to correlate with coal cycling events and because of the system's larger gas-fired generation capacity. Identifying days when wind generation resulted in the cycling of coal units allowed for a precise understanding of the emission impacts. The gravity and frequency of these events increased as more wind generation was introduced to the system. This mirrors the results found on the PSCO system, supporting the theory that increased rates of cycling arise from the incremental integration of wind generation. Furthermore, these wind-driven, coal-cycling events resulted in significantly more SO_2 and NO_X emissions than if wind generation had not been utilized. The same results were found on the PSCO system. Not only does wind generation not allow ERCOT utilities to decrease SO_2, NO_X, and CO_2 emissions, it is directly responsible for creating more SO_2 and NO_X emissions and CO_2 emission savings are minimal at best.

General Conclusions and Future Outlook

Our studies detail the surprising conclusion that the use of wind energy in the PSCO and ERCOT contexts results in increased emission rates for SO_2 and NO_X and, in the case of PSCO, CO_2. The mechanism driving increased rates is the need to cycle coal facilities to accommodate wind—a must-take resource under the respective states' RPS mandates. When wind generation comes online, generation from coal (and natural gas-fired plants is curtailed until the wind subsides, then nonwind generation is again ramped up to meet demand. Cycling coal units in this manner drives their heat rates up and their operating efficiencies down, emitting more SO_2, NO_X, and CO_2 than would have been emitted if the units had not been cycled.

Two caveats must be understood when interpreting these results. First, we found no instances in which PSCO violated any of its air permits as a result of cycling coal. Neither PSCO case study indicated that PSCO's emissions exceeded its permits. Furthermore, the study authors are not suggesting that PSCO violated its permits in extrapolating the case study results to estimate annual emissions. The second caveat pertains to the data. For the ERCOT analysis, hourly generation data by plant and fuel type including wind was available. Thus, it was possible to precisely identify wind events based on a sudden decline in coal generation coupled with a simultaneous increase in wind generation. For PSCO's territory, it was not possible to define wind events with the same precision because PSCO does not release hourly generation data for its wind resources. Subsidiary conclusions based on our analysis include:

Duration — Cycling coal-fired power plants has short- and long-term impacts. Studies of the interactions of coal and wind often mention the cycling issue, but generally discuss the impacts in a very narrow context—the duration during which the coal plant reduces generation. This study concludes that the impacts frequently have much longer durations. Many instances were found where cycling caused bag houses and other pollution controls to lose their calibration and take as long as 12 to 15 hours, sometimes as long as 24 hours, to settle back to pre-event emission rates. During these periods, emission rates normally exceeded what would be experienced if the plants ran at stable generation levels.

Timing — Wind-induced coal-plant cycling appears to be a night-time phenomenon. Nearly 70% of the cycling instances identified for PSCO in 2008 occurred between 12:00 and 8:00 a.m. Similarly, 82% of coal cycling events at ERCOT occurred at the same time of night.

Nonwind renewable implications — Coal-cycling issues do not appear to impact solar and other nonwind renewable energy forms. Solar energy is generated during daylight, thus coinciding with natural gas-fired generation. When solar energy peaks, the likelihood is much greater that natural gas-fired generation can be cycled to accommodate the energy.

Generation mix — Composition of the generation stack is a critical factor. Most wind-driven cycling events appear to occur between 12:00 and 8:00 a.m.—during periods of lowest load. As a result, PSCO and ERCOT utilities operate only their baseload facilities then. In the PSCO context, this means the coal plants supplemented with some combined-cycle natural gas and hydro are in operation. ERCOT's baseload includes nuclear, coal, and combined-cycle plants. The extra emissions result because the RPS-mandated must-take wind resources exceed the quantity of power generated from combined cycle gas. PSCO's generation mix from 12:00 to 8:00 a.m. averages 62% coal, 20% combined cycle, and 18% hydro, wind, and purchases. ERCOT's corresponding mix is 17% nuclear, 40% coal, 28% combined cycle, 6% combustion turbine, 9% wind, and 0% hydro. Increasing the proportion of baseload generated by more flexible generation equipment such as natural gas-fired combined cycle plants and stored energy sources will enable systems to absorb wind without having to cycle their coal plants.

Regulatory conflict — The study results suggest that the RPS mandate is in conflict with the Colorado State Implementation Plan for air emissions. The RPS standard requires that more wind resources be utilized than can be offset with lower-emission, natural gas generation equipment. That is the case today when wind resources account for about 9% of PSCO's total sales. Wind generation will increase in the coming years due to mandates to move toward a 30% of total sales standard. Without substantially more natural gas generation added to the PSCO system, the emission increases documented in this study will rise, further enlarging the degree to which Denver and the Front Range violate the State Implementation Plan limitations.

National implications — Congress is considering legislation that would mandate a federal RPS. While our study paid only cursory attention to areas beyond the ERCOT and PSCO territories, it is doubtful that a national RPS can be imposed without creating the same emissions outcome found in the

ERCOT and PSCO territories and in many other states. Unless other states have sufficient natural gas "cushions"—Texas has the largest share of its generating capacity fueled by natural gas—imposition of an RPS greater than 5% will probably increase emissions of CO_2, NO_X, and SO_2.

The results of this study should not be interpreted as a critique of wind energy. Rather, they suggest modifications to its development to ensure that the benefits are more fully realized. Current RPS standards mandate that wind is a must-take resource and other generation facilities must be cycled to accommodate it. If utilities have sufficient variable generation facilities online to avoid cycling their coal units, these RPS provisions work well. However, where these conditions are lacking, alternative approaches are required to avoid cycling coal plants. The alternatives might include construction of additional variable generation sources such as natural gas and stored energy systems, the imposition of additional emission controls on coal units to make them more flexible, or changing dispatch orders. Whatever the approach, care must be taken that imposition of the RPS does not exacerbate ozone and other air quality issues.

References

1. Bentek Energy. April 16, 2020. How less became more: wind power and unintended consequences in the Colorado energy market. Evergreen, CO.
2. Zavadil, R. May 2006. Wind Integration Study for Xcel Energy/Public Service Company of Colorado. Prepared by EnerNex Corporation, p. 47. http://www.nrel.gov/wind/systemsintegration/pdfs/colorado_public_service_windintegstudy.pdf
3. Zavadil, R. December 2008. Wind Integration Study for Xcel Energy/Public Service Company of Colorado. Addendum: Detailed Analysis of 20% Wind Penetration. Prepared by EnerNex Corporation. Appendix B.
4. Antoine, M., T. Matsko, and P. Immonen. 2000. Modeling Predictive Control and Optimization Improves Plant Efficiency and Lowers Emissions. ABB Power Systems; Telesz, R. November 2000. Retrofitting Lime Spray Dryers at Public Service Company of Colorado. Babcock & Wilcox; both presented at PowerGen International, November 14–16, 2000.
5. Vierstra, S., and D. Early. 1998. Balancing Low NO_2 Burner Air Flows through Use of Individual Burner Airflow Monitors. AMC Power; presented at PowerGen International, December 9–11, 1998.
6. PSCO. (2008, 2010). Wind Generation in PSCo Commercial Operations. Retrieved from XcelEnergy.com: http://www.xcelenergy.com/sitecollectiondocuments/docs/CRPExhibit2PSCOIntegratedReliabilityTraining.pdf

3

Pumped Hydroelectric Energy Storage

Jonah G. Levine

CONTENTS

Basic Concepts

Pumped hydroelectric energy storage (PHES) technology allows utilities to store generated energy that may become a load on demand within the constraints of the particular facility. Energy is stored as the potential of water raised against gravity. Stated simply, water is pumped through a turbine from a lower reservoir (afterbay) to a higher reservoir (forebay) and the activity uses energy. When it is desirable to have that energy or water returned, the water is allowed to flow back through a turbine from the higher reservoir (forebay) to the lower reservoir (afterbay) and vice versa. Figure 3.1 is a simple line drawing of a PHES facility.

Value of PHES to Interconnected Electricity Systems

As the world embarks on a new era of distributed and at times nondispatchable electrical energy systems, the ability to manage increased levels of

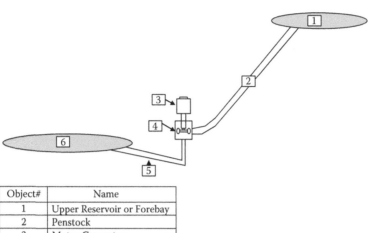

Object#	Name
1	Upper Reservoir or Forebay
2	Penstock
3	Motor Generator
4	Pump Turbine
5	Tail Race
6	Lower Reservoir or Afterbay

FIGURE 3.1
Line drawing of PHES facility.

variable generation will be critical to every operating region. The electric grid of tomorrow must exhibit flexibility in load and generation management. The world's generation mix is becoming more diverse and variable. This variability is one of the costs for decreased emissions and increased fossil fuel savings. How can utility companies and other energy providers manage increased variability? Each operating region will have to assess the resources available and address the variability inherent in its system. Effective management to ensure load and generation flexibility will require a series of steps including:

- Improving energy efficiency and implementing demand response
- Utilizing spatial and source generation diversity exhibiting complementary profiles
- Bringing resources to market via transmission and timely utilization
- Energy storage
- Improving electric utility data communications to integrate the above steps

Of the five steps for providing flexibility to an electricity system energy, storage is a key step for accommodating variability. A PHES facilitates the alignment of renewable generation with loads. Baseload generation has the largest impact on emissions factors and produces most of the energy. If renewable generation is to impact electricity-driven emissions in a significant way, it must be able to affect baseload generation. When renewable generation comes online coincident with low demand, the challenge is to ramp down the baseload thermal generation systems by some method of curtailment—and ramping down is not always possible. Storage, specifically PHES, can address the difficulty in managing ramping rates and anti-correlation of variable generation and loads.

PHES takes energy from the grid and returns it at a later time when it is needed. This raises an important question. What resource powers the PHES? The resource that powers a PHES facility will be the at the margin when the PHES facility pumps. Thus, when coal is on the margin, coal will power the PHES facility and wind will furnish the power when wind is on the margin. The greater the amount of renewable energy on the system, the greater the possibility for having a renewable energy resource on the margin as the prime mover for the PHES facility. The larger the percentage of renewable energy resource on the system, the more flexibility it will require.

At lower penetrations of renewable resources, less storage is needed and the probability is greater that the storage will be powered by nonrenewable energy resources. At higher penetrations of renewable resources, more storage will be required and the probability is greater that the storage will be powered by renewable resources. Thus, storage will become critical to further development of renewable energy and will reflect the emissions reductions as the system lowers its overall emissions.

Example: Dominion Power's Bath County PHES Station

Dominion Power has released a video covering many aspects of its Bath County pumped storage facility. For a basic discussion of pumped storage stations, the video may be accessed online.[1] A brief listing of facts about the Bath County facility is as follows:

Net generating capacity	2,100 MW
License issued	January 1977
Start of commercial operation	December 1985
Cost (1985)	$1.7 billion or $810/MW
Owners	Dominion Power (60%), Allegheny Power (40%)
Lower dam	135 feet (41 meters) high; 2,400 feet (732 meters) long; contains 4 million cubic yards (3.1 million cubic meters) earth and rock fill
Lower reservoir	555 surface acres (2.25 square kilometers); water level fluctuates 60 feet (18 meters) during operation
Upper dam	460 feet (140 meters) high; 2,200 feet (671 meters) long; contains 18 million cubic yards (13.8 million cubic meters) of earth and rock fill
Upper reservoir	265 surface acres (1.07 square kilometers); water level fluctuates 105 feet (32 meters) during operation

Water flow, pumping	11 million gallons/minute (694 cubic meters/second)
Water flow, generating	14.5 million gallons/minute (915 cubic meters/second)
Turbine generators	Six Francis type 350 MW units manufactured by Allis Chalmers
Maximum pumping power per unit	563,400 horsepower (420,127 kw)

PHES Efficiency

The process of pumping water up and releasing it back down to achieve the return of energy is not 100% efficient. Some of the electric energy used to pump the water up will not be returned as usable electric energy on the way back down. This efficiency loss is incurred as a result of rolling resistance and turbulence in the penstock and tail race and efficiency losses in the motor generator and pump turbine. In addition, the water retains some energy as it flows into the tail race. Considering all of these losses, PHES has a turnaround efficiency ranging from 70 to 80%, dependent on design characteristics. For example if a PHES facility were 80% efficient, that would mean for every ten units of energy put into storage, eight can be returned on demand. Table 3.1 reflects PHES cycle efficiencies for plants constructed after the late 1970s.[2]

Facilities in United States

Figure 3.2 is a map showing PHES facilities in the United States. Table 3.2 lists all PHES facilities (by state) as reported by the 2005 EPA EGRID.

Energy and Power Potential

PHES facilities require two fundamental resources: elevation change (head) and water. By ascertaining the potential elevation change and water available, it is possible to determine the power and energy availability of a PHES facility with the basics of gravitational potential energy or the fluid power equation:

$$PE = mgH \tag{3.1}$$

where PE = potential energy in joules, m = mass [volume (m^3) · density 1000 kg/m^3], g = acceleration due to gravity or 9.81 m/s^2; and H = hydraulic head height in meters (m).

TABLE 3.1

PHES Cycling Efficiency

	Low %	High %
Generating Components		
Water conductors	97.40	98.50
Pump turbine	91.50	92.00
Generator motor	98.50	99.00
Transformer	99.50	99.70
Subtotal	87.35	89.44
Pumping Components		
Water conductors	97.60	98.50
Pump turbine	91.60	92.50
Generator motor	98.70	99.00
Transformer	99.50	99.80
Subtotal	87.80	90.02
Operational	98.00	99.50
Total	**75.15%**	**80.12%**

Source: Chen, H.H. 1993. Pumped storage. In *Davis' Handbook of Applied Hydraulics*, 4th ed. Zipparro, V.J. and H. Hansen, Eds. McGraw Hill, New York. 22.23.

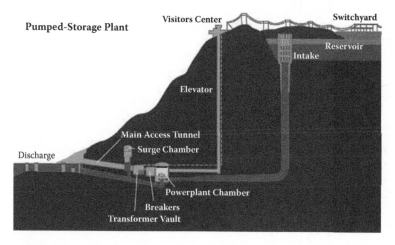

FIGURE 3.2
Map of PHES facilities in United States.

TABLE 3.2

United States PHES Facilities as Reported by 2005 EPA EGRID

State Abbreviation	Plant Name	Plant Operator Name	Parent Company Name Associated with the Operator	Plant Latitude	Plant Longitude	Number of Generators	Plant Capacity Factor	Plant Nameplate Capacity (MW)	Plant Annual Net Generation (MWh)
PSTATABB	PNAME	OPRNAME	OPPRNAMES	LAT	LON	NUMGEN	CAPFAC	NAMEPCAP	PLNGENAN
AR	Degray	USCE-Vickburg District	US Army Corp of Engineers	34.0575	−93.1714	2	0.0997	68.0	59,402.0
AZ	Horse Mesa	Salt River Project		33.3596	−112.4878	4	0.0556	129.5	63,065.0
AZ	Mormon Flat	Salt River Project		33.3596	−112.4878	2	0.0490	63.5	27,229.0
AZ	Waddell	Central Arizona Water Conservation Dist		33.3596	−112.4878	4	0.1531	40.0	53,644.9
CA	Castaic	Los Angeles City of		34.5198	−118.6062	7	0.0254	1,331.0	295,809.0
CA	Edward C Hyatt	California Department of Water Resources		39.6618	−121.5917	6	0.3243	644.1	1,829,689.0
CA	Helms Pumped Storage	Pacific Gas & Electric Co	PG&E Corp	36.7548	−119.6397	3	−0.0097	1,053.0	−89,046.0
CA	J S Eastwood	Southern California Edison Co	Edison International	36.7548	−119.6397	1	0.1935	199.8	338,715.0
CA	ONeill	USBR-Mid Pacific Region	US Bureau of Reclamation	37.1869	−120.7037	6	−0.1448	25.2	−31,958.0
CA	Thermalito	California Department of Water Resources		39.6618	−121.5917	4	0.2450	115.1	247,006.0

CA	W R Gianelli	California Department of Water Resources	37.1869	-120.7037	8	-0.1052	424.0	-390,893.0
CO	Cabin Creek	Public Service Co of Colorado	39.6856	-105.6370	2	-0.0299	300.0	-78,446.0
CO	Flatiron	USBR-Great Plains Region	40.6650	-105.4607	3	0.2212	94.5	183,097.0
CO	Mount Elbert	USBR-Great Plains Region	39.1970	-106.3409	2	-0.0537	200.0	-94,116.0
CT	Rocky River	Energy Capital Partners' First Light	41.7926	-73.2249	3	0.0487	31.0	13,213.0
GA	Carters	USCE-Mobile District	34.7885	-84.7453	4	0.1255	500.0	549,578.0
GA	Richard B Russell	USCE-Savannah District	34.1156	-82.8419	8	0.1050	628.0	577,373.0
GA	Rocky Mountain Hydro	Oglethorpe Power Corporation	34.3500	-85.3036	3	-0.0643	847.8	-477,761.0
GA	Wallace Dam	Georgia Power Co	33.2722	-82.9986	6	0.0064	321.2	17,885.0
MA	Bear Swamp Brookfield Power USA	Brookfield Asset Management Inc (Canada)	42.3693	-73.2013	2	-0.0252	600.0	-132,611.0
MA	Northfield Mountain	Energy Capital Partners' First Light	42.6123	-72.4458	4	-0.0400	940.0	-329,032.0
MI	Ludington	Consumers Energy Co	43.9984	-86.2520	6	-0.0638	1,978.8	-1,106,241.0
MO	Clarence Cannon	USCE-St Louis District	39.5288	-91.5284	2	0.1309	58.0	66,501.9

(Continued)

TABLE 3.2 (CONTINUED)

United States PHES Facilities as Reported by 2005 EPA EGRID

State Abbreviation	Plant Name	Plant Operator Name	Parent Company Name Associated with the Operator	Plant Latitude	Plant Longitude	Number of Generators	Plant Capacity Factor	Plant Nameplate Capacity (MW)	Plant Annual Net Generation (MWh)
PSTATABB	PNAME	OPRNAME	OPPRNAMES	LAT	LON	NUMGEN	CAPFAC	NAMEPCAP	PLNGENAN
MO	Harry Truman	USCE-Kansas City District	US Army Corp of Engineers	38.2941	−93.2915	6	0.1987	161.4	280,881.0
MO	Taum Sauk	AmerenUE	Ameren Corp	37.3636	−90.9764	2	−0.0665	408.0	−237,594.0
NC	Hiwassee Dam	Tennessee Valley Authority		35.1331	−84.0589	2	0.2001	165.6	290,278.0
NJ	Yards Creek	Jersey Central Power&Light Co	FirstEnergy Corp	40.8488	−75.0004	3	−0.0712	453.0	−282,707.0
NY	Blenheim Gilboa	New York Power Authority		42.4442	−74.4419	4	−0.0491	1,000.0	−429,914.0
NY	Lewiston Niagara	New York Power Authority		43.2015	−78.7430	12	−0.1669	240.0	−350,817.0
OK	Salina	Grand River Dam Authority		36.3073	−95.2319	6	−0.0610	288.0	−153,825.0
PA	Muddy Run	Exelon Energy	Exelon Corp	40.0457	−76.2523	8	−0.0663	800.0	−464,490.0
PA	Seneca	FirstEnergy Generation Corp	FirstEnergy Corp	41.8160	−79.2795	3	−0.0600	469.0	−246,551.0
SC	Bad Creek	Duke Carolinas LLC	Duke Energy	34.9599	−82.9185	4	−0.0641	1,065.2	−598,001.0

SC	Fairfield Pumped Storage	South Carolina Electric&Gas Co	SCANA Corp	34.3899	-81.1164	8	-0.0760	511.2	-340,525.0
SC	Jocassee	Duke Carolinas LLC	Duke Energy	34.8831	-82.7233	4	-0.0485	612.0	-260,149.0
TN	Raccoon Mountain	Tennessee Valley Authority		35.0471	-85.3975	4	-0.0446	1,530.0	-597,935.0
VA	Bath County	Dominion Virginia Power	Dominion	38.1937	-79.8099	6	-0.0451	2,100.6	-829,353.0
VA	Smith Mountain	Appalachian Power Co	American Electric Power Co	36.9927	-79.8773	5	-0.0145	547.5	-69,472.0
WA	Grand Coulee	USBR-Pacific NW Region	US Bureau of Reclamation	47.9555	-118.9849	33	0.3433	6,809.0	20,474,048.0

Map displaying a sampling of pumped storage plants in the United States.

$$P = Q \times H \times \rho \times g \times \eta \tag{3.2}$$

where P = generated output power in watts (W), Q = fluid flow in cubic meters per second (m³/s), H = hydraulic head height in meters (m), ρ = fluid density in kilograms per cubic meter (kg/m³) = 1000 (kg/m³) for water, g = acceleration due to gravity (m/s²) or 9.81 (m/s²), and η = efficiency.

As shown in Equation (3.2), the changeable variables are the volumetric flow, the head, and the facility efficiency. Assuming that the head and a great deal of the efficiency will be dictated by the location of the facility, the flow becomes a significant design point of a PHES facility. The head and the flow have an important relationship: if the head is larger, the water utilization can be minimized. The reverse is also true: if the flow can be maximized, the head can be reduced. Examples of PHES facilities that use the tradeoffs between head and flow can be found. The facility in Ludington, Michigan, maximizes flow to accommodate a moderate head.[3] In locations with high head and limited water, it would be desirable to maximize head to reduce the water needed. The water needed in a PHES facility is not consumed; it is reused (less losses for evaporation and seepage) by multiple up and down pumping. In many cases, the water is releasable back into the system when needed.

As seen in Equation (3.2), to derive the energy of a facility one must input the volumetric flow of water. One way to derive the appropriate flow is to assess the potential sizing for the limiting reservoir. Ascertain the total volume of water available in the limiting reservoir. Assess the energy market to which the PHES facility will be interconnected and ascertain a desirable energy storage time for that market. Knowing the volume of fluid and the time desired for storage, one can suggest a flow rate by dividing the storage volume by the storage time desired. The flow rate may also be dictated by the economics of penstock construction or limitations of current or developable waterways.

Continuing to assess the tradeoffs between flow and head requires assessment of the total energy available in the raised volume of water. By ascertaining the energy available, one can dictate the dispatch of that energy by varying the power of the specific system. For a simple rule of thumb, assume that for each acre foot of water raised 1 meter, 3 kWh are stored for later dispatch or, for each acre foot of water raised 1200 feet, 1 MWh of energy is stored for later dispatch. Figure 3.3 depicts energy potential over the volume of water needed at increasing heads.

Development

Sites for PHES development are challenging to locate but this should not infer that no sites are available for development. Alternative development approaches should be considered. In addition, one should consider current

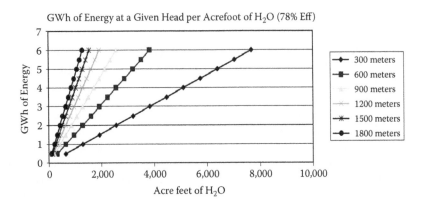

FIGURE 3.3
Energy potential over volume of water necessary at increasing heads; plots beyond 600 m approach or surpass current technology.

resources on the electric grid. It may be the case that the electric operation areas needing storage resources have pump-back capacities that are either not dispatched as pump-back or run solely as peaking resources. As greater variability is injected into the electric grid based on increased variable generation systems and changing loads, it may benefit a system to redispatch current resources in a different manner.

The two basic requirements for a PHES facility are elevation change (head) and water. Alternative designs can make the challenges of water or head availability less difficult to overcome. Head availability is traditionally provided by above-ground elevation change. Where above-ground elevation change cannot be found, it may be possible to use the elevation change between the ground surface and a location below the surface of the earth, thus the head constitutes the difference between the surface elevation and a below-ground reservoir. This may be feasible in a mine or possibly in line with pumped water between a subsurface aquifer and a surface reservoir.

Society may also gain the advantage of dispatchable water pumping to align the movement of water with the availability of energy; this strategy combines pumped storage and demand response. The primary hurdle of aligning timed water pumping and availability is transparent and regular communication between those who plan and operate water systems and those who plan and operate energy systems. Where water resources are not plentiful, it may be possible to utilize creative planning or alternative design to provide the needed water resources.

Examples of alternative water design include the coupling of an agricultural water supply with a PHES facility and the utilization of water produced by the gas and oil extraction industry. In the latter case, the produced water would have to be cleaned to an acceptable environmental standard to avoid the distribution of organic and inorganic pollutants contained in the water. It is possible that the revenue from a PHES facility could pay for the cleanup of produced water.[4] These alternatives for attainment of head and water should be considered and will likely be required to make some future projects work.

Environmental Considerations

Any hydro-based development project will involve environmental considerations. Preserving healthy aquatic environments is important. By bringing multiple stakeholders into the development conversation early, challenges may be avoided down the road.

Environmental considerations key to development include maximizing head to reduce the water need; potentially utilizing alternative water sources and cleaning those sources if needed; constructing PHES facilities off line to avoid damming of running rivers. Environmental considerations must be assessed on a case-by-case basis but should be seen as solvable challenges that can provide environmental benefits if addressed by multiple stakeholders early in the planning process.

System Components

Reservoirs

The reservoirs serving a PHES system are critical to its viability. The reservoirs known as the forebay and afterbay are the storage tanks of the system. They are critical from both technical design perspectives and may be "show stoppers" from social and environmental perspectives. The difficulties in siting reservoirs warrant utilizing current reservoirs where available as opposed to developing new ones. Finding suitable reservoirs already in place may be possible but new reservoirs will likely be required for future development. The higher the available head, the smaller the reservoirs must be based on the tradeoff between head and flow for power and energy availability.

Upper Reservoirs

New forebay development can be accomplished by various ways including a stream valley reservoir (Figure 3.4) or a hill top reservoir. A stream valley reservoir is created by an impoundment constructed across a stream valley such that it fills the valley behind the impoundment. A derivation on stream valley construction common to PHES development is a high stream valley reservoir that follows the same basic idea of valley impoundment but applies it to a steeper slope. An example of a high stream valley forebay is Cabin Creek located just outside Georgetown, Colorado, owned and operated by Xcel Energy/Public Service Company of Colorado (Figure 3.5).

A hilltop reservoir is constructed by building an embankment around a hilltop and storing water inside the embanked hilltop. An example of a hilltop reservoir is the forebay of Raccoon Mountain (Figure 3.6).

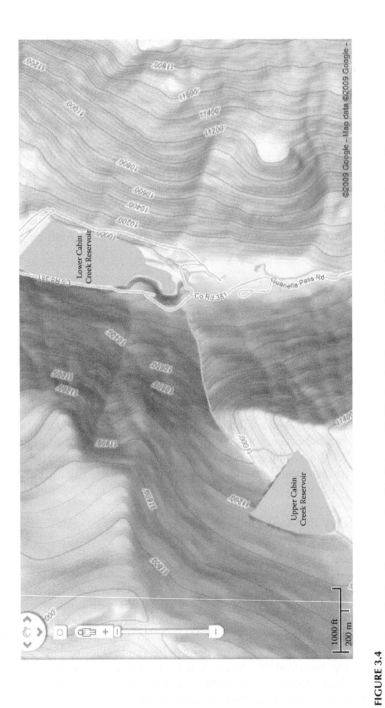

FIGURE 3.4
Topographic view of high valley reservoir of Xcel Energy/Public Service Company of Colorado's Cabin Creek operation.

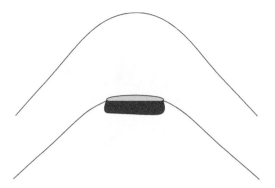

FIGURE 3.5
Simple line drawing of valley reservoir; design is also applicable to high valley reservoir.

An upper reservoir must have a design element to deal with floods (natural inflows) and overpumping. A spillway is designed to allow overflow waters to vacate the reservoir without damaging it. One serious failure due to overpumping and inability to drain without damage occurred at the Taum Sauk pumped storage project. The Federal Energy Regulatory Commission (FERC) incident description covering the dam breach appears below.

Incident Description: Taum Sauk Pumped Storage Project (No. 2277). On December 14, 2005, at approximately 5:20 a.m. CST the northwest corner of the Taum Sauk Pumped Storage Project No. 2277 upper reservoir rim dike failed, resulting in a release of the upper reservoir. The reservoir was reported to have drained in about one-half hour. Approximately 4,300 acre feet of storage water released. The breach flow passed into East Fork of Black River (the river upstream of the lower Taum Sauk Dam) through a state park and campground area and into the lower reservoir. The Lower Taum Sauk Dam was reported to be overtopped and did not sustain damage. Upon leaving the Lower Taum Sauk Dam area, the high flows proceeded downstream of the Black River to the town of Lesterville, Missouri, located about 3.5 miles downstream from the lower dam. The incremental rise in the river level was about 2 feet which remained within the banks of the river.[5]

On December 14, 2005, the project's upper reservoir breached, rendering the facility inoperable. The upper reservoir overtopped when the pumps filling it failed to shut off. Erosion undercut the rock fill dam, creating a breach that emptied the reservoir.[6] To avoid overpumping challenges, Kermit Paul proposed the following recommendations[7]:

- PHES facilities should have a fail-safe overpumping safety design.
- This design should be independent of water level control systems, as water level and monitoring control and overpumping protection are separate systems.

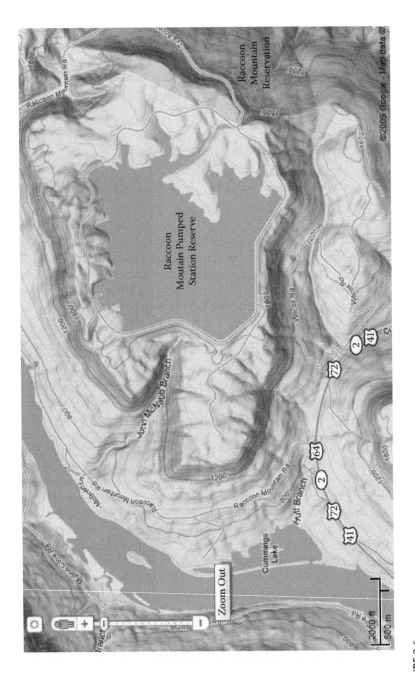

FIGURE 3.6
Topographic view of mountain-cut ring dike, upper reservoir.

- This design should have direct action pump shutdown.
- This system should be redundant.
- This system should have mechanisms for testing and calibration.

Lower Reservoirs

Lower reservoirs or afterbays may be found on existing reservoirs or in stream and river valleys. The area of the afterbay should be large enough to accommodate the spillage needs of the forebay. Alternative site designs for an afterbay may include oceans, large lakes, various underground configurations, water treatment ponds, and agricultural water storage reservoirs.

Waterways

The elements of the waterways needed for a PHES system are the headworks, penstock, tailrace, and one or more surge tanks that allow the water to be moved between the forebay and the afterbay.

Headworks

The headworks connect the forebay with the penstock and serve as the entrance in the generation mode and the exit in the pumping mode. This dual directional requirement of a PHES facility means that the flow must avoid vortices in both directions to maximize efficiency of the facility. The head works should also include trash racks to remove debris and prevent it from entering the system.

Penstocks

A penstock or main water conduit between the forebay and the turbo machinery is an important design component of a PHES facility. A PHES may have single or multiple penstocks located above or below ground.

A major siting consideration for a PHES facility is minimizing the ratio of total water conductor length to the head. Ideally this ratio is 1:1 where the total conductor length is equal to the head; thus the positioning of the forebay would be directly overhead of the turbo machinery (motor or generator). See Figure 3.7.

When sizing a penstock for a Francis or propeller-type turbine more than 5 feet in diameter one can start with the empirical formula developed by Sarkaria.[8] With an approximate penstock diameter and a desired power rating, one can approximate water velocities.

$$D = 4.44 \ (P^{0.43}/H^{0.65}) \tag{3.3}$$

where D = economical diameter of penstock (feet), P = rated horsepower (hp) of turbine, and H= rated head of turbine (feet).

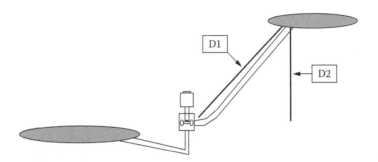

FIGURE 3.7
Basic concept of using ratio of total conduit distance to head distance. The ratio should be less than or equal to 10:1; 1:1 is optimal. D1 is the conductor distance and D2 is the head distance.

Water conduit sizing has a direct connection to the starting time of a PHES unit. The starting time for a unit should be between 1 and 2.4 seconds and not exceed 2.5 seconds.[9] The water starting time of a hydro facility is the time required for the water to move through the conduits to the turbo machinery. This may be calculated by summing the lengths of constant diameter sections of conduits times the velocity of the fluid in that section, over a gravitational constant times the net head. This calculation[9] is expressed as

$$T_W = \sum LV/gh \qquad (3.4)$$

where T_w = water starting time, L = length of constant diameter section of water conduit, V = average flow velocity in a related section of L, g = gravitational constant, and h = net head. Mechanical starting time is the time required for the turbo machinery to reach rotational speed and begin electrical operation.[9]

$$Tm = (WR^2 \times n^2)/(1{,}620{,}000 \times P) \qquad (3.5)$$

where T_m = mechanical starting time, WR^2 = product of weight of revolving parts (shaft turbine runner and generator rotor) and the square of their radius of gyration, n = rotational speed of turbine and generator for a direct connected synchronous generator, and P = turbine full gate capacity (hp). The mechanical starting time over the water starting time (T_m/T_w) measures unit stability. If a PHES unit is expected to follow load or variable generation for integration and provide frequency regulation, the ratio of T_m/T_w should be equal to or greater than 5.[9]

Draft Tube

Where a reaction turbine is used, a draft tube is necessary and is designed simultaneously with the pump turbine. The draft tube takes water from the turbine runner and into the tailrace below the elevation of the tail water. The tube allows the full head of the plant to be used because it facilitates the utilization of suction head.

Tailrace

A tailrace is a water conduit between the afterbay or tail water and the draft tube or turbo machinery. A tailrace conveys water from the tail water during pumping and into the tail water during generation.

Surge Tank or Chamber

Surge tanks can be provided both upstream and downstream of a water conduction system.[10] The purpose of a surge tank is to dampen changes in pressure, protecting the water conduits, turbine, and pumping equipment. A surge tank allows the turbo generator to regulate its load. Figure 3.8 shows surge chambers on the upstream and downstream sides of turbo machinery.

Iwabuchi et al.[11] show that optimizing governor operation allows the sizing of surge chambers to be minimized, thus leading to lower development cost. Since surge chambers or tanks provide dampening mechanisms to water conductors and mechanical equipment, it is logical to further develop these devices to furnish more flexibility for PHES operations. A research group in Austria is pursuing that via a project titled: "Design of Pumped Storage Schemes."[12] A brief description follows.

The objective is the development of a new surge tank system for pumped storage schemes (PSS) to govern the changing requirements of electricity networks arising from the integration of renewable, volatile energy sources. The novel design provides for splitting the lower chamber of a two-chamber surge tank. The two separate portions of the lower chamber are situated at different levels and connected via an overflow sill at the lower end of the riser. Likewise, the water column is separated under the critical loading conditions, and thus can accelerate the water column in the tailrace tunnel while simultaneously building up the required back pressure for the pump.

Turbo Machinery

Most PHES developments are designed with Francis style pump turbines. These devices both pump and generate; they are generally categorized as reaction turbines (Figure 3.9). Another option is the use of a separate pump and turbine. This design may allow a project to utilize higher heads and maintain higher technical efficiency during both generation and pumping operations. Although higher technical efficiency can be achieved, the economic penalty must be assessed against the financial benefit from the efficiency gain. This system also allows the selection of an impulse turbine and a centrifugal pump.

Figure 3.10 displays hydraulic reaction turbines at operational head ranges versus turbine discharge.[13] Power output increases with head and flow up

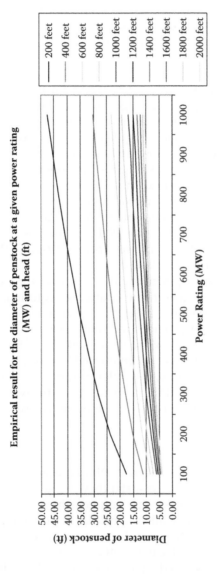

FIGURE 3.8
Empirical result for diameter of penstock at given power rating and head.

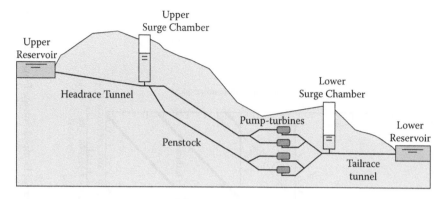

FIGURE 3.9
Upper and lower surge chambers of PHES design. (From Garrity, J. J et al. 1985. In *Handbook of Energy Systems Engineering Production and Utilization*, Wiley Interscience: New York. With permission.)

the y-axis and from left to right across the x-axis. Above the head range of a reaction turbine, an impulse turbine is usable. Table 3.3 displays pump turbine head limits.

Figure 3.11 depicts limits and ratings of a system developed by Rodrique[14] and in operation since 1979. Data for single stage and multiple stage reversible pump turbines and tandem units with impulse and Francis turbines are shown.

Impulse Turbines and Centrifugal Pumps

Very high head applications may utilize impulse turbines with separate centrifugal pumps. Impulse turbines function by running the energy contained in the raised water through a nozzle where the water becomes a jet and is directed at the vanes or buckets of a wheel. The jet of water hitting the vane is the impulse that removes the kinetic energy from the fluid and transfers it to the turbine or wheel. An impulse turbine cannot pump and thus must be matched to a separate centrifugal pump. Centrifugal pump sizing is covered in a U.S. Bureau of Reclamation monograph.[15]

Reaction Turbines

Most modern PHES facilities utilize reaction turbines. These turbines function by reactive forces. Unlike the arrangements of impulse turbines, the nozzles or buckets of reaction turbines are attached to the scroll cases. As a scroll case turns, the water either is pushed by the machine or pushes the machine, based on the operating status—pumping or generating, respectively. Reversible pump turbines fall into three subgroups: radial flow Francis type, mixed flow diagonal, and axial flow.[16]

Radial flow Francis type turbines — Also known as Francis style pump turbines, these types are the most common applications for PHES facilities.

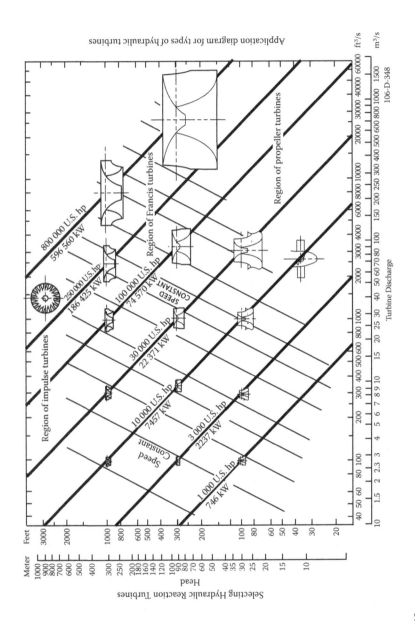

FIGURE 3.10

Reaction turbine sizing relative to head and flow. (From Iwabuchi, K. et al. 2006. Advanced Governor Controller for Pumped-Storage Power Plant and Its Simulation Tool. SICE-ICASE International Joint Conference. Korea. *IEEE Explore*, pp. 6064–6068. With permission.)

TABLE 3.3

Pump Turbine Head Limits

	Facility	Turbine		Pump		Manufacturer(s)
		Max Head (m)	Max Power (MW)	Max Head (m)	Max Power (MW)	
Single-stage reversible pump turbines	Ohira (Japan)	512	256	545	269	Hitachi, Toshiba
	Raccoon Mtn (US)	316	400	323	400	Allis Chalmers
	Bajina Basta (Serbia)	600	315	621	310	Toshiba
	Bath County (US)	384	457	387	420	Allis Chalmers
Multistage reversible pump turbines	La Coche (France)	930	79	944	80.6	Neyrpic, Vevey
	Edolo (Italy)	1224	122	1237	142	Hydroart, De Pretto Escher Wyss
	Chiotas (Italy)	1047	147	1069	160	Hydroart, De Pretto Escher Wyss
Tandem units with impulse turbines	San Fiorano (Italy)	1401	140	1439	106	De Pretto Escher Wyss
	Rottau (Austria)	1100	200	1100	144	Voith
Tandem units with Francis turbines	Homburg (Germany)	625	243	625	250	Voith, Escher Wyss
	Rosshag (Austria)	672	58	736	59	Voith, Escher Wyss

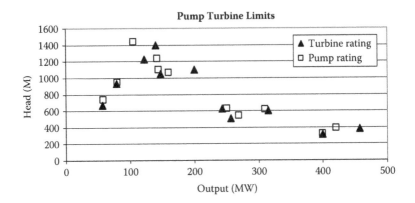

FIGURE 3.11
Pump turbine limits and ratings (head versus output). (From Iwabuchi, K. et al. 2006. Advanced Governor Controller for Pumped-Storage Power Plant and Its Simulation Tool. SICE-ICASE International Joint Conference. Korea. *IEEE Explore,* pp. 6064–6068.)

They are capable of handling a wide range of heads—75 to 1300 feet (20 to 400 meters).

Mixed or diagonal flow turbines — These turbines are typically applied between 35 and 300 feet (10 to 90 meters) This design is smaller and more agile than the design of a radial flow Francis type. Mixed or diagonal flow units have faster pump starting times and adapt to variability more easily than Francis units. Despite their smaller size, these units typically cost more than Francis types.

Axial flow pump turbines — These turbines are applied between 3 and 45 feet (1 to 15 meters) of head. Their low head application is well suited for tidal operations. Axial flow turbines exhibit reasonable efficiency within the full spectrum of their operating ranges.

Variable Speed Turbines

PHES facilitates are regularly characterized by high variations in head efficiency and thus losses are incurred during operation. Optimal design points for the pumping and generating modes differ. To address these challenges, dual speed units use an arrangement of stator windings to switch between numbers of poles. The two settings achieve synchronous speeds.

While dual speed units have only begun to address the varied operating conditions of PHES facilities, a recent development can enable variable speed units. Karl Scherer describes the opportunities and challenges of variable speed units for PHES.[17] These units function by using double-fed motor generators that operate as asynchronous units by feeding low frequency alternative current into excitation windings. Variable speed units achieve higher generation efficiencies over head ranges and allow variable power consumption during pumping operation.

References

1. Bath County Pumped Storage Station. *Mountain of Power*. YouTube: http://www.youtube.com/watch?v=mMvOZSVXlzI (accessed November 2010).
2. Chen, H.H. 1993. Pumped storage. In *Davis' Handbook of Applied Hydraulics*, 4th ed. Zipparro, V.J., and H. Hansen, Eds. McGraw Hill, New York.
3. http://www.consumersenergy.com/welcome.htm?/content/hiermenugrid.aspx?id=31 (Ludington, Michigan PHES).
4. Levine, J., and F. Barnes. 2010. Energy variability and produced water: two challenges, one synergistic solution. *ASCE Journal of Engineering* 136 (March 2010): 6–10.
5. Federal Energy Regulatory Commission. July 2008. Taum Sauk Report. http://www.ferc.gov/industries/hydropower/safety/projects/taum-sauk.asp
6. Federal Energy Regulatory Commission. December 2007. Order Granting Intervention, Denying Rehearing, and Dismissing Request for Stay. Project No. 2277-005. http://ferc.gov/whats-new/comm-meet/2007/122007/H-4.pdf (accessed November 2010).
7. Paul, Kermit. November 2006. Overpumping protection systems design criteria. http://www.ferc.gov/industries/hydropower/safety/initiatives/november-workshop/over-pumping-protection.pdf
8. Sarkaria, G. S. 1979. Economic Penstock Diameters: A 20-Year Review, *Water Power and Dam Construction*, Vol. *31*, No.11. United Kingdom.
9. Hansen, H., and Antonopoulos, G.C. 1993. Hydroelectric plants. In *Davis' Handbook of Applied Hydraulics*, 4th ed. Zipparro, V.J., and H. Hansen, Eds. McGraw Hill, New York.
10. Garrity, J.J. et al. 1985. Hydroelectric power. In *Handbook of Energy Systems Engineering Production and Utilization*, Wilbur, L., Ed. Wiley Interscience, New York, p. 1142.
11. Iwabuchi, K. et al. 2006. Advanced governor controller for pumped storage power plant and its simulation tool. *Proceedings of SICE-ICASE International Joint Conference*. Korea. IEEE, Washington, p. 6064.
12. http://www.energiesystemederzukunft.at/results.html/id4344 (Energy systems of tomorrow 2007: design of pumped storage schemes).
13. U.S. Department of the Interior, Bureau of Reclamation. 1976. *Selecting Hydraulic Reaction Turbines*. Monograph 20. http://www.usbr.gov/pmts/hydraulics_lab/pubs/EM/EM20.pdf
14. Rodrique, P. 1979. The selection of high head pump turbine equipment for underground pumped hydro application. Pump turbine schemes: planning, design, and operation. ASME-CSME Applied Mechanics Fluid Engineering and Bioengineering Conference. Buffalo, NY.
15. Duncan, W., and C. Bates. 1978. *Selecting Large Pumping Units*. U.S. Bureau of Reclamation. Engineering Monograph 40. http://www.usbr.gov/pmts/hydraulics_lab/pubs/EM/EM40.pdf
16. Sun, J. 1993. Hydraulic machinery. In *Davis' Handbook of Applied Hydraulics*, 4th ed. Zipparro, V.J., and H. Hansen, Eds. McGraw Hill, New York.
17. Scherer, K. 2005. Change of speed. 13th International Seminar on Hydropower Plants. http://www.waterpowermagazine.com/story.asp?storyCode=2027383

4

Underground Pumped Hydroelectric Energy Storage

Gregory G. Martin

CONTENTS

Introduction

Underground pumped hydroelectric energy storage (UPHES) is an adaptation of conventional surface-pumped hydroelectric that uses an underground cavern or water structure as a lower reservoir. Conceptually, this seems a logical and sound solution to energy storage. The practical design and actual construction of large UPHES systems represent challenging tasks. Smaller UPHES systems may be more easily built, especially if existing underground and surface structures can be used. To date, no UPHES system, commercial or otherwise, has ever been installed and used. This chapter will overview concepts and design considerations for UPHES systems, both large and small.

The lower reservoir is the heart of the UPHES system. It can be excavated from suitable geologic rock at various depths or it can tap an existing aquifer or other naturally occurring underground water containment. UPHES alleviates several of the challenges encountered with surface-pumped hydroelectric installations. Dependence on surface topology is eliminated, although suitable underground geology and structures are required. An underground system has a vertical water flow path that greatly reduces loss associated with transverse water flow. The environmental impact of an underground installation is smaller than those of conventional pumped hydro systems because only one surface reservoir is required. UPHES systems eliminate new river dams and large powerhouses on the surface, minimize wildlife habitat disruption, and reduce noise. Figure 4.1 illustrates a basic large, excavated UPHES system.

System Sizing

UPHES systems are roughly classified as small (10 kW to 0.5 MW) and large (0.5 to 3000 MW) installations. Large systems are usually targeted at mitigating the varying loads of major urban centers or providing buffers to make variable renewable energy more consistent. Most studies performed for large

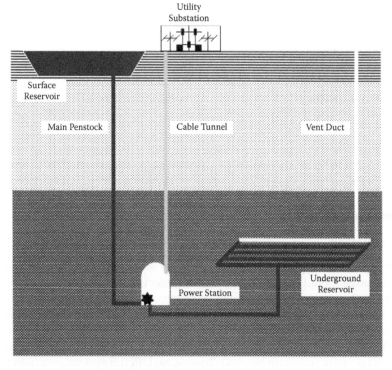

FIGURE 4.1
Large basic UPHES system.

UPHES have considered installations of between 1000 MW and 3000 MW as the most economical scale, and have analyzed the economics based on a varying consumer load supplied by conventional coal-fired power plants.

Smaller installations serve single users, small communities, agriculture, or industrial operations. These smaller installations can utilize an existing underground water structure and consume no net water (except for evaporation). The economical sizing of a smaller UPHES is a complicated matter; cost of electricity, geological formations, water table characteristics, existing infrastructure, user load profiles, and renewable energy source availability all contribute to optimal sizing of the system.

Design Overview

Generally, a large system requires excavation of the lower reservoir from suitable geologic strata. Most studies of UPHES have targeted hard rock (such as granite) lower reservoir beds for favorable structural properties. Solution mining of underground salt domes to create a large underground caverns has also been proposed. A large lower reservoir would be excavated in a network of extruded, narrow caverns rather than a single large cavern. This method improves the structural integrity of reservoir excavations and is better suited to known excavation techniques. The system can be designed for single- or double-drop configuration. A single-drop configuration is shown in Figure 4.1, and a double-drop installation utilizes an intermediate storage reservoir at half the depth of the main reservoir.

The availability and capability of water turbines and pumps that can operate at very high pressures contribute to the decision of how deep a lower reservoir is built and whether to use a double-drop system. However, a design trade-off must be considered for reservoir depth because as depth increases, the volume of water required to generate the same power decreases.

Equation (4.1) below is the basic power calculation for UPHES and illustrates the trade-off between depth and water flow.

In addition, large scale UPHES systems require the main power station to be located below the lower reservoir to eliminate cavitation issues that could reduce the lifetime of the machinery. An underground power station calls for safe personnel access to a great depth below the surface. The obvious major barrier to deployment of UPHES is the difficulty of excavating a large, stable water reservoir at a significant depth below the earth's surface, presumably from hard subterranean rock. An underground power station is not an amenable location for human occupation and must be operated remotely.

$$P = Q \cdot H \cdot \rho \cdot g \cdot \eta \tag{4.1}$$

where P = power generated in watts [W]. If horsepower [hp] is used in the equation, all other variables' units must be changed accordingly. Q = fluid flow in cubic meters per second [m³/s], ρ = water density in kilograms per

cubic meter [kg/m³], H = hydraulic head height in meters [m], g = acceleration due to gravity [m/s²], and η = efficiency.

The study of small scale applications of UPHES is even less common
than studies of large scale uses. The UPHES literature is essentially devoid
of research on small scale (hundreds of kilowatts) installations. It is possible that, under certain conditions, a small scale UPHES system may make
economic sense and provide added benefits for utilizing variable renewable
energy sources. In cases where water is used extensively for crop irrigation,
modifying an existing irrigation and well infrastructure to accommodate a
small UPHES system may be feasible. Subsequent sections of this chapter
give an in-depth review of such a system known as aquifer UPHES.

Literature Review

Research and studies of UPHES systems are not numerous. Since the incep-
tion of the commercial concept of UPHES in the 1970s, relatively little research
has been done. Most of the literature and studies surfaced in the late 1970s
and early 1980s, followed by a drought of academic, government, and private
interest in the topic. Most UPHES studies focused on large, gigawatt scale
(1000 MW) systems. The concept has been studied mainly as a method to
match variable load demands of large urban centers to the constant power
outputs of coal or nuclear plants.

A large UPHES system is a major undertaking requiring careful planning, long term financing, advanced machinery, and large scale excavation.
Uncertainty about the structural integrity and detailed layouts required for
underground hard rock structures further complicate the plant design and
planning. For these reasons, historical studies that analyzed specific sites for
UPHES have not inspired the funding and massive effort required to build
a large system.

A good overview of the economics and challenges of UPHES appears in a
1978 Aerospace Sciences Meeting paper[1] that addressed the basic concepts and
layouts of large UPHES systems as well as expected economical sizing. Tam,
Blomquist, and Kartsouns of the Argonne National Laboratory presented a
review of UPHES status, technologies, and market in 1979.[2] Allen, Doughtry,
and Kannberg of The Pacific Northwest Laboratory studied the concept in
the early 1980s.[3] Findings from these research efforts suggest that the economical size of a UPHES system is in the range of 1000 to 3000 MW, suitable
for large urban areas of about a million people or more. Both reports call for
modernized turbo machinery capable of higher head (pressure) operations,
further studies of underground cavern geology, and system optimization.

A 1981 U.S. patent granted to James L. Ramer of Waukesha, Wisconsin,
claimed intellectual property over the concept of using UPHES with

an underground salt dome.[4] Several plant cycles are used to dissolve underground salt (a process sometimes called solution mining) until a large underground cavern is attained. The plant then continues operation as a hydro storage and generation resource.

Far earlier in patent history, a U.S. patent application for a "System of Storing Power" was filed on June 7, 1907, by R. A. Fessenden and a patent (No. 1,247,520) was granted on November 20, 1917. In it, Fessenden stated:

> The invention herein described relates to the utilization of intermittent sources of power and more particularly to natural intermittent sources, such as solar radiation and wind power, and has for its object the efficient and practical storage of power so derived....
>
> It has long been recognized that mankind must, in the near future, be faced by a shortage of power unless some means were devised for storing power derived from the intermittent sources of nature....
>
> These sources are, however, intermittent and the problem of storing them in a practicable way, i.e., at a cost which should be less than that of direct generation from coal, has for many years engaged the attention of the most eminent engineers, among whom may be mentioned Edison, Lord Kelvin, Ayrton, Perry, and Brush.

Fessenden went on to describe possible methods of storing energy by moving water from one elevation to another. Realization of the need for an efficient and practical method of storing energy, for easy conversion to and from electricity, is not new. For more than 100 years, people have searched for efficient ways to store and deploy electrical energy. With the advent of modern renewable energy generation, the need for large scale energy storage is even more urgent, and advances in water pumps, turbines, and excavation techniques bring the concept closer to reality.

A site analysis study was performed for a large UPHES installation in the state of Illinois in 1982. It focused on site selection, tunneling layout, machinery, logistics, and costs.[5] The project was never funded for construction. In more recent history, a detailed geologic analysis of a proposed large UPHES installation was performed by Uddin in 2003.[6] His study focused on structural integrity analysis of excavating a lower reservoir in subterranean hard rock, and brought to light the challenges of excavating reservoirs for deep underground storage of water.

Small (Aquifer) UPHES

This section introduces, describes, and analyzes an aquifer underground pumped hydroelectric energy storage system. Aquifer UPHES is a new adaptation of underground pumped hydroelectric energy storage that uses an underground aquifer as the lower reservoir.[7] The basis of this concept is the utilization of gravitational potential energy in surface water with respect to an aquifer or water table below the surface. The proposed system design, operation, necessary technologies and components, and aquifer characteristics are described.

System Description and Operation

The aquifer underground pumped hydroelectric storage system is designed to store energy in the form of gravitational potential energy in water separated between a surface reservoir and a subterranean aquifer. Energy is stored by pumping water from the underground source into a surface reservoir for storage. This energy is later recovered by releasing surface-stored water back to the source through a turbine that generates electricity. Figure 4.2 illustrates an aquifer UPHES system. This storage system is most suitable when used in tandem with variable renewable energy sources such as wind and solar photovoltaics because the energy storage serves to buffer the variable output and reliably supply on-demand power for user loads. The main elements of the aquifer UPHES system include:

- Source of electricity (solar panels, wind turbine, grid)
- Surface reservoir or pond
- Deep, high flow capacity water well
- Integrated motor-pump turbine generator unit
- Electrical center (power electronics, controls, protection)

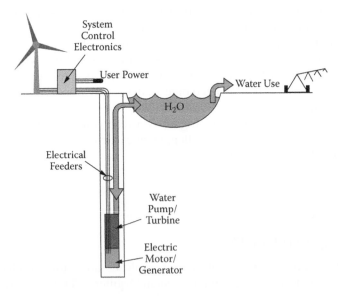

FIGURE 4.2
Aquifer UPHES system.

The system is designed to maximize the power output capability of an installation. To this end, the efficiency of the turbine, the available hydraulic head, and the flow capability of the well are maximized. To calculate the power output during the generation cycle, the basic fluid power Equation (4.1) applies.

Figure 4.3 plots various hydraulic head values on a power versus flow plane, assuming a turbine efficiency of 70%. The calculation was performed using metric units and the results were converted to English units. The plot shows that power output is maximized when hydraulic head and flow are also maximized. While the head is generally dictated by the characteristics of the installation site, the flow parameter can potentially be increased, as discussed in the following sections.

In addition to power output, it is desirable to maximize the energy output delivered by the system. Energy storage capacity is determined by the volume of stored water and the rated power (head, flow, and efficiency) of the system:

$$\text{Energy [kWh]} = \text{Power [kW]} \cdot \text{Time [h]} \qquad (4.2)$$

It is desirable to maximize hydraulic head to achieve the maximum possible energy output. Flow and reservoir volume are closely coupled parameters that affect the energy capacity and are constrained by the required duration

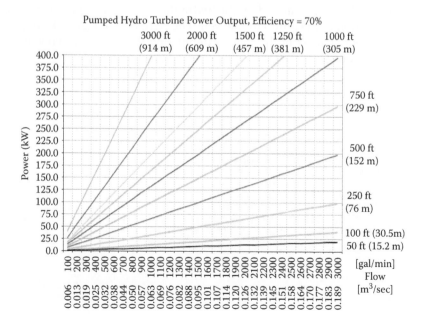

FIGURE 4.3
Relationship of power to flow and hydraulic head.

of power generation. These parameters will be determined by both site characteristics and the end use requirements.

Performance Modeling

The most important parameters for optimization in the design of this system
are the well hydraulic head, flow capacity, and electrical system efficiency.
Contrary to common well flow yield measurements, the parameter of interest here is the measured flow that can be re-injected into the aquifer, not
the flow that can be pumped out or "yielded." While aquifer re-injection is
accomplished at various projects across the country, methods to accurately
determine re-injection flow capacity are more complicated than the common
pumping calculations. This section will provide simplified models to predict
the re-injection flow of a well with a given hydraulic head. It will also analyze
the allocation of hydraulic head between the head that powers the turbine
and the head that re-injects water into the aquifer. Also addressed are the
electrical system performance and efficiency during electricity generation.

The initial thought of a designer of this type of system is to approximate
the re-injection flow capacity as roughly the same as the yield capacity of
a well. Let us test this assumption for steady state flow conditions. When
drawing water from a well, a cone of depression is created around the well
because of the finite transmissivity of the aquifer material. This cone can
depress down to the point at which the pump is located and no further.
Thus, the well yield is limited by the hydraulic conductivity of the material
and the location of the pump in the well. The right half of Figure 4.4 depicts
two-dimensional effect of the cone of depression that occurs when water is
drawn from a well.

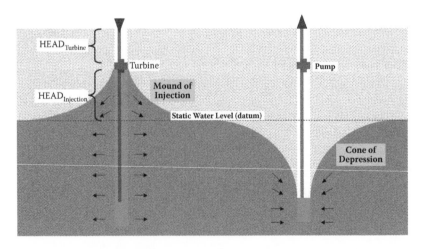

FIGURE 4.4
Mound of injection and cone of depression.

When water is injected into a well, the opposite phenomenon (mound of injection) occurs, also as a result of the finite hydraulic conductivity of the aquifer material. The left half of Figure 4.4 illustrates the mound of injection. The injection flow rate depends on hydraulic head and transmissivities of the aquifer. The turbine or pump location affects the injection flow rate insofar as it drops some of the hydraulic head that could function to "push" more water into the aquifer at a higher rate. That is, there is a trade-off between the amount of head allocated to the turbine for electricity generation and the amount of head pressure functioning to inject water flow back into the aquifer.

The governing equation describing hydraulic and water flow parameter interactions is the corollary in groundwater hydraulics to the thermal conduction problem. The general form of the groundwater equation for water flow in an aquifer[8] is

$$\frac{S}{T} \cdot \frac{\partial h}{\partial t} = \frac{1}{r} \cdot \frac{\partial}{\partial r}\left(r \cdot \frac{\partial h}{\partial r}\right) \tag{4.3}$$

where S = storage coefficient, T = transmissivity [ft²/min] (m²/s), h = hydraulic head [ft] ([m]), drawdown = $h_{initial} - h$ [ft] ([m]), and r = radius from well [ft] ([m]).

This equation applies to confined aquifers. However, the drawdown or head calculated for a confined aquifer using this equation can be correlated to the height of an injection mound operating in an unconfined aquifer. If the assumption that the pumping occurs over a long time is adopted, the Cooper-Jacob approximation to the Theis equation, expressed in terms of drawdown over time, can be used:

$$drawdown = \frac{2.3 \cdot Q}{4 \cdot \pi \cdot T} \cdot \log\left(\frac{2.25 \cdot T \cdot t}{r^2 \cdot S}\right) \tag{4.4}$$

where Q = water flow [ft³/min] [m³/s]. With the goal of estimating the height of the mound of injection (negative drawdown), the following assumptions are made:

Q = –133.7 ft³/min (–0.0631 m³/s)

S = 0.1 (unconfined aquifer) or 0.0001 (confined aquifer)

T = 2 ft²/min (0.00308 m²/s)

r = 1 ft (0.305 m)

t = 6 hours = 360 min

k_T = T ÷ aquifer thickness (ft/min or cm/s)

Solving for well drawdown at r = 1 ft (0.305 m):

$$drawdown = \frac{2.3 \cdot (-133.7 \frac{ft^3}{min})}{4 \cdot \pi \cdot 2 \frac{ft^2}{min}} \cdot \log\left(\frac{2.25 \cdot 2 \frac{ft^2}{min} \cdot 360\,min}{1\,ft^2 \cdot 0.1}\right)$$

$$drawdown = 51.5\,ft = 15.7m$$

This result, that the mound of injection rises 51.5 feet (15.7 meters) above the water table, presents a difficulty. The well is only 200 feet (61 meters) deep, so the water rises 25% of the way up the well when water is injected in this fashion. The storage coefficient (S) has a much smaller effect on the mound height, although an aquifer having a high S experiences a decrease in drawdown or a decrease in mound height. On the other hand, if the transmissivity is raised to 10 ft²/min (154.8 cm²/s), the mound height decreases to 12 feet (3.7 meters). Therefore, transmissivity plays an important role in the height of the injection mound and in the design of the power output of the system.

This analysis indicates that an aquifer with high transmissivity (high hydraulic conductivity) is needed. Figure 4.5 shows a plot of mound height versus transmissivity, holding the other values given above constant. This plot demonstrates the trend of decreased mound height as transmissivity increases.

Based on this analysis, the original hypothesis of whether the same flow that can be yielded by pumping can be re-injected into a well has been tested.

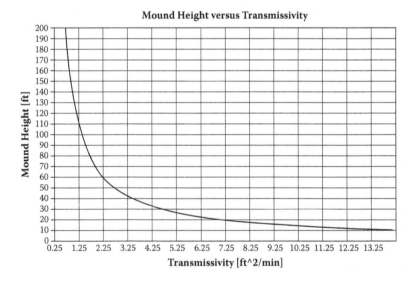

FIGURE 4.5
Mound height versus aquifer transmissivity.

The result depends on the aquifer transmissivity and the depth to water. In many cases, the same flow that can be pumped out can indeed be re-injected, but the hydraulic head available for turbine operation is reduced. In an aquifer with low transmissivity, injection results in a mound that can potentially reach the surface. In comparing this to the pumping cycle, if the pump depth below the water datum level is the same as the depth from the surface to the water datum level, the mound of injection will just reach the surface, negating the ability to produce power from the injection flow using a turbine.

This modeling exercise indicates that the aquifer UPHES system must be designed with aquifer transmissivity, injection mound height, and depth to water as major design parameters. The transmissivity must be relatively large so that the mound of injection remains low enough to reserve sufficient hydraulic head for turbine power generation. In the case of the 200-foot (61-meter) water depth example, if the transmissivity of the aquifer is 6.5 ft²/ min (100.6 cm²/s), then there remain 182 feet (45.7 meters) of head for turbine operation.

This situation can be modeled as a simple electrical circuit with a voltage source representing the total hydraulic head potential and resistances representing the "head drop" for the turbine and for the aquifer injection mound. The current in the circuit represents water flow. The resistance associated with the turbine correlates to the resistance to water flow in the pipe and in the turbine. The injection resistance correlates to the transmissivity (resistance to water flow) encountered in the aquifer. Figure 4.6 shows an electric circuit model for system head. This electrical model gives accurate insight into the interactions of the design parameters. Holding the total head constant, a transmissivity increase correlates to a reduction in the injection resistance ($R_{Injection}$ in the figure). A reduction in the injection resistance is coupled with an increase in the turbine resistance, keeping total flow constant, but increasing the power output of the turbine. Alternately, decreasing the flow while holding total head and transmissivity constant will decrease the injection head. Then, more head (voltage) will drop across the turbine.

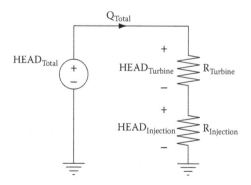

FIGURE 4.6
Electrical circuit model for hydraulic head.

Alas, because the flow has been decreased, the total power dissipation from the turbine will remain constant. Using the correlation between Ohm's law and equations governing the flow, head, and hydraulic resistance, design trade-offs can be calculated (Figure 4.6 and Equation [4.5]). The basic equations governing the linear behavior of an electric circuit (Ohm's Law) are

$$HEAD_{Total} = HEAD_{Turbine} + HEAD_{Injection}$$

$$HEAD_{Turbine} = Q_{Total} * R_{Turbine}$$

$$HEAD_{Injection} = Q_{Total} * R_{Injection} \qquad (4.5)$$

$$POWER_{Turbine} = Q_{Total} * HEAD_{Turbine}$$

Given the resistances associated with turbine piping and aquifer hydraulics, the relative trade-off among flow, head, and power output can be modeled using Equation (4.5). Also, a correlation can be derived relating the transmissivity in the hydraulic circuit to the resistance in the equivalent electric circuit. The allowable flow (current) through the circuit is proportional to the transmissivity (resistance) in the circuit. The equations to determine the resistance and conductance in an electrical circuit are

$$R = \frac{l}{\sigma \cdot A}; \quad G = \frac{1}{R} = \frac{\sigma \cdot A}{l} \qquad (4.6)$$

where R = resistance [Ω], G = conductance [S], l = length [m], σ = conductivity [S/m], and A = area [m²]. The analogous equation in hydraulics for transmissivity is

$$T = \frac{k \cdot A}{r}$$

$$T = k \cdot b \qquad (4.7)$$

$$T = \frac{\kappa \cdot \gamma \cdot b}{\mu}$$

where T = transmissivity [m²/s] or [ft²/min], k = hydraulic conductivity [m/s] or [ft/min], A = area [m²] or [ft²], r = radius [m] or [ft], b = aquifer thickness [m] or [ft], κ = intrinsic permeability [m²] or [ft²], γ = specific weight of water [1000 kg/m³ (at 4°C)], and μ = dynamic viscosity of water, [0.00089 Pa/s]. Comparison of these two equations shows that electrical conductance is the corollary to transmissivity, and electrical conductivity is the corollary to hydraulic conductivity. Transmissivity has units of area per time, and is usually calculated as the hydraulic conductivity times the aquifer thickness.

The next step would be to derive an expression for the transmissivity in an aquifer experiencing recharge flow through a well. Although not included here, the result of this derivation should match the governing equation above for hydraulic flow in an aquifer.

Transmissivity (T) is a measure of the volume of water flowing through a cross-sectional area of an aquifer [for example, 1 foot times the aquifer thickness (b)] under a hydraulic gradient (for example 1 ft /1 ft) in a given amount of time. Transmissivity is a parameter used to calculate water flow in aquifers, and is equal to hydraulic conductivity (k) times aquifer thickness (b), as shown in Equation (4.7). Hydraulic conductivity (and therefore transmissivity) depends on the permeability of the medium, specific weight of water, and dynamic viscosity of water. The equation for hydraulic conductivity is found from application of Darcy's law.

Transmissivity therefore depends on the above quantities, including another length dimension, aquifer thickness. In this text, the quantity of transmissivity is used to evaluate water flow and aquifer performance. It should be noted that transmissivity can be related back to the basic properties of aquifer materials. Another common material property, porosity, is the ratio of the empty space volume to the total volume in a material. Porosity can change with depth, because the weight of material from above compresses the voids between particles. While porosity of a material can affect the intrinsic permeability, these quantities are not necessarily related.

Water Pump Turbine

The core of the aquifer UPHES system is an integrated pump turbine and motor generator unit. As the name suggests, this single unit performs the functions of both pumping water using electrical power and generating electricity from water power. This type of integrated machine exists commercially for large pumped hydroelectric installations, normally employing a Francis reaction type turbine coupled to a synchronous AC electric machine. A unit sized and designed for the proposed aquifer UPHES application is not yet commercially available. In this section, the important design considerations for the integrated pump turbine and motor generator unit for use in an aquifer UPHES system are described.

An option for the design of the aquifer UPHES pump turbine is the use of standard centrifugal or "vertical turbine" well pumps in the forward direction for pumping and in reverse for turbine operation.[9] Figure 4.7 shows an example of a submersible vertical turbine pump. This use of the device is known as a pump-as-turbine (PAT) design. A first order estimation of the turbine efficiency of a centrifugal pump is that it is the same as the pump efficiency. Although originally designed as a pump, a centrifugal pump may

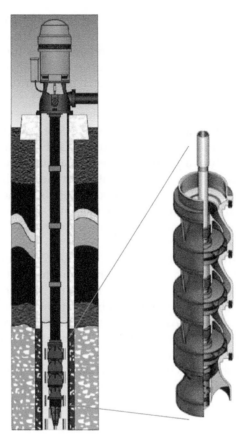

FIGURE 4.7
Example of submersible vertical turbine pump. (Figure courtesy of American Turbine Vertical Turbine and Submersible Pumps. http://americanturbine.net/sites/americanturbine.net/files/brochures/vertical-turbine-submersible-pump-brochure.pdf)

be capable of operating in reverse as a turbine at efficiencies in the range of 65 to perhaps 85%.[10] This method is proposed as a preferred option for the aquifer UPHES situation because it uses existing technology, is commercially available, and represents a low cost solution. Because of the difficulty in predicting turbine performance of a specific centrifugal pump, testing is required to characterize the flow capability, water velocity range, and turbine efficiency. The selected centrifugal pump design must employ a keyed shaft to accommodate shaft torque in either direction.

Centrifugal motor pumps are commonly used for pumping water in many situations. They are available in submerged or nonsubmerged designs, with a wide range of available head ratings, flow ratings, and power ratings for commercial versions. These units are commonly centrifugal or vertical turbine designs, integrated with AC induction motors. The industry standard estimation of pump efficiency is 55%, but with proper system design, a centrifugal pump could achieve as much as 85% efficiency. The efficiency for the pumping cycle or the turbine cycle can be optimized, but they cannot be optimized simultaneously. For aquifer UPHES, the turbine efficiency must be optimized. The assumed ranges of machine efficiency used in this text are 70 to 85% for turbine operation and 65 to 80% for pump operation. These numbers are estimates adopted from data on modern PAT pumps, centrifugal pumps, and turbines.

Reaction type turbines, such as Kaplan or Francis designs, are capable of accomplishing both pumping and turbine functions at efficiencies that increase with unit size. Kaplan or propeller style turbines are used in low head, high flow applications. Francis turbines and PAT designs are applied in high head, high flow situations. Typical efficiencies for very large Francis turbines can approach 95%.[10] For smaller units, lower efficiencies in the range of 70 to 90%, depending on head, flow, and specific speed, can be expected. In standard Francis designs, the water enters or leaves a scroll-shaped vane housing at a right angle to the rotation of the drive shaft. This characteristic may pose a design challenge in installing such a unit in a vertical shaft well.

Electric Motor Generator

Motor generator units that operate with relatively high efficiency represent a mature and available technology. As with pump turbines, the efficiency increases with size and rating. Some large motor generators can operate above 96% efficiency. For the application in question, efficiencies between 88 and 94% may be achievable.

Well motor pumps commonly employ AC induction motors or synchronous wound-rotor AC motors for larger machines. Although commercial designs assume the unit will be used as a motor only, modifications can produce efficient operation in generator mode In the case of a synchronous wound-rotor AC machine, an interface to the machine's rotor windings that provides excitation current during generating is needed. This modification is relatively simple to implement with or without power electronics. If power electronic control of the winding is used, the frequency, and therefore the speed and torque, of the generator output can be regulated.

For AC induction machines, perhaps the most common and simple modification involves connecting excitation capacitors to the three-phase leads of the machine.[11] These capacitors provide excitation current that is 90 degrees out of phase with the primary generation current waveform. This excitation current induces currents in the rotor of the machine that allow it to operate as a generator.

Another possible modification involves the use of power electronics to synthesize the excitation current. The same electronics used to drive the machine as a motor are used to control excitation while it is generating. To

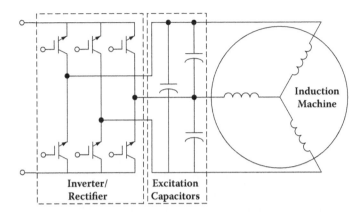

FIGURE 4.8
Induction machine connections for generator modifications.

implement this method, a more complicated control loop is programmed into the machine controller software. Figure 4.8 schematic indicates the connections of the excitation capacitors and the basic inverter/rectifier power electronic switches.

For an aquifer UPHES system, a centrifugal well pump with an induction motor is recommended. This option represents the lowest cost solution, but efficiency during the generating cycle may not be optimized. For the final system design, a full-sized unit should be procured and tested to determine the actual performance capabilities. Care must be taken to select a unit that will operate with the required flow and range of attainable water velocities.

Electrical System

An electrical system is needed to implement the aquifer UPHES function and interface it with energy sources, user loads, and the utility grid. Its main functions include:

- Power electronics motor drive to energize motor pump during the pumping cycle
- Generator exciter and rectifier to extract electricity from the turbine generator during the generating cycle
- Grid tie inverter to condition the power to 60 Hz, 480 Vac; includes rectifier function in the case of a local wind turbine power source
- A 480 Vac circuit breaker panel for protective functions and power routing
- A transformer to convert 480 Vac to 220 Vac and 120 Vac for user load power

- A 220 Vac and 120 Vac circuit breaker panel for protection and power routing to user loads
- A system control, monitoring, and user interface panel that regulates and controls the entire system

Figure 4.9 shows a block diagram of the connections of an example electrical system. All the components introduced above and shown in the figure are described in more detail below. In reality, a system would likely only have one local renewable energy source (solar panels or wind turbines). Also, it is possible that this system could be run off the grid, but emergency back-up power provisions such as batteries may be required. In general, the functional components (with the exception of the system controller) are available commercially. Further detailed engineering design work is required to correctly interface and control these components in a concerted and safe manner.

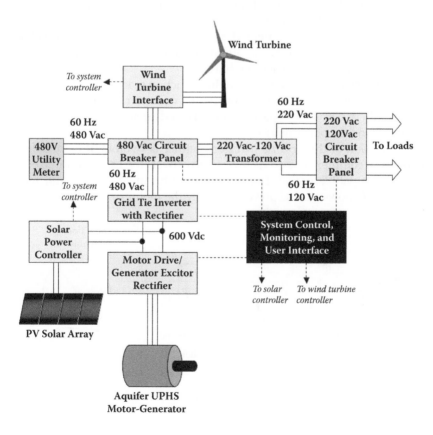

FIGURE 4.9
Electrical system block diagram.

A power electronics controller is desired to interface the motor generator to the system. This controller has two main functions. It must electrically drive the motor during pumping operation. This involves inverting the DC voltage using a pulse width-modulated, six-step, trapezoidal or other motor drive strategy to control a three-phase power electronics-based inverter. The impedance and voltage drop in the long lines between the inverter and the motor (located near the bottom of the well) must be taken into account. This inverter could be designed to drive the motor at only a single speed (simpler implementation) or at variable speeds. A variable speed drive has the ability to use lower power input (such as when the solar or wind source is minimal), thus increasing the efficiency of the pumping cycle. Additionally, it is possible to further optimize the pumping cycle by matching the photovoltaic solar voltage and current to the pump characteristic using a method such as maximum power point tracking (MPPT).[12]

The controller must excite the motor generator and rectify the output. Two methods of exciting the generator were discussed earlier. It is recommended here to employ the scheme involving advanced control of the power electronics switches to simultaneously excite the machine and rectify the output. The excitation capacitors are eliminated, reducing cost and increasing reliability. Figure 4.10 is a schematic of the proposed unit, utilizing position feedback sensed directly from the machine shaft.

A filter must be employed between the induction machine and the inverter. This filter attenuates the voltage spikes that occur on the lines due to their long length. A DC link capacitor is connected to stiffen the DC bus and improve transient performance.

The function of the grid tie inverter/rectifier is twofold. It is intended to operate as a commercial grid tie inverter to convert the DC power into 60 Hz, AC grid-compatible power. In addition, the unit must step up and rectify incoming AC power to DC power to supply the motor drive controller.

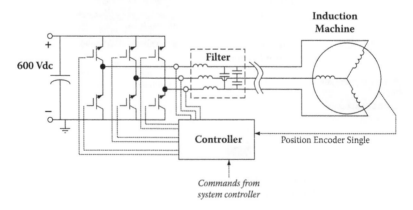

FIGURE 4.10
Motor control inverter/rectifier.

Circuit breaker panels that protect and control the AC systems are required. The circuit breakers should be implemented using appropriate relays or contactors so that power routing can be accomplished by the system controller. These relays also function as protective circuit interrupt elements. In the case of tying the grid inverter to the utility meter, the system controller must monitor and verify that the frequency and voltage waveforms are compatible with the grid. At that point, the system controller will close the circuit breaker connecting the system to the grid.

A transformer with its primary winding connected to the 480 Vac system is utilized to provide 60 Hz AC power to the low voltage circuit breaker panel. This panel houses either traditional passive circuit breakers for the user loads or externally controlled relays (or contactors) if additional automation is desired.

The system controller is responsible for the overall control and protection of all the other elements of the electric system. It has several important functions, with its primary job to appropriately route power to or from the storage system, the local power sources, and the loads. To implement energy storage management, the controller monitors the amount of power generated by the local power sources, the load demand power present, and estimates the status of the energy storage system (full, empty, 50% full, etc.). Based on this information, it initiates one of the following actions:

- If energy is being generated but not used by the loads, power is routed to the motor pump drive and water is pumped to the surface, until the surface reservoir is full.

- If power is demanded by the loads, but no power is generated, stored energy is released by putting the storage system into generating mode, until storage reserve is deleted.

- If power is demanded by loads, the energy storage is depleted, and no local power generation is online, electricity is routed from the utility grid to supply the load demand.

- If more power is produced by the local energy source than is being used and the storage reservoir is full, power will be "net metered" or routed to the grid.

Direct energy from the utility grid can also be stored. This option would be used if time-of-day pricing of grid electricity is in effect. For example, if less expensive grid electricity is available at night than during the day, this inexpensive electricity can be stored and later used when grid prices rise. The efficiency of the storage system must be traded against the cost differential of the time-of-day pricing to determine whether this choice is economically beneficial.

In addition to energy storage management, the system controller performs monitoring, protection, and power routing functions. System status including which circuit breakers are closed and open, which units are operating and in which direction, power flow data, and other parameters are continually monitored. Each individual electrical system component has provisions for self protection against overloads and overheating, but the job of the system controller is to ensure that no system configuration that may damage equipment is enabled.

The user interface to the operation of the overall system is housed in the system controller. This interface tells the user the status of the system, including the output from local power sources, the status of the energy storage, the loads that are energized, and the power flow specifics. The interface also allows the user to configure the system in certain ways and shut down components or sections.

Water Well

An aquifer UPHES system utilizes a deep, high flow capacity water well to accomplish energy storage. The power capacity of the turbine is a function of available head and flow. Analysis from previous sections shows how the trade-off of head, flow, mound height, and aquifer transmissivity affects the system design. In this section, well characteristics are reviewed and methods to increase system power by modifying wells or using infiltration pits are described.

Water well characteristics vary greatly in installations across the world and also within the same aquifer system. The main characteristics of importance to the aquifer UPHES system are

- Transmissivity or hydraulic conductivity of the surrounding geologic formation
- Depth to water
- Well diameter
- Well casing
- Confined or unconfined aquifer

As an example, a well having greater than 1000 gal/min (0.063 m^3/s) injection flow capacity and 300 feet (91 meters) of head for turbine power generation is targeted. To achieve this, an aquifer with 350 feet (107 meters) depth to water must have transmissivity of about 2.6 ft^2/min (40.3 cm^2/s) or greater. Does this type of well exist? What can be done to retrofit a well to achieve the necessary parameters?

Because typical irrigation wells have limited flow capacity, it follows that one would consider ways to increase the injection flow capability of a well to extract the maximum possible power. Aquifer recharge (AR) and aquifer storage and recovery (ASR) wells are examples of systems designed to re-inject water into an aquifer. To quote one source, "Currently, more than 60 aquifer storage and recovery (ASR) sites are in operation around the U.S. These projects range from a single well to networks of 30 wells, with recovery capacities ranging from 500,000 gallons per day from single wells to 100 million gallons per day from well fields."[13] Recharge wells are designed to replace water in an aquifer or underground structure by flowing water backward into a well, thereby recharging the aquifer. In ASR wells, water is both injected and removed, depending on seasonal cycles and water use obligations. This type of well sets the precedent for an aquifer UPHES installation, although aquifer UPHES cycles are much more frequent. Modified AR and ASR wells designed for direct injection operations have been proposed.[14] Figure 4.11 and Figure 4.12 show options for modified wells to increase recharge flow capacity. ASR wells can be cost-effective and can be easily integrated with existing water utility facilities using well fields.

Essentially, these concepts serve to increase possible injection flow by increasing the completed surface area in contact with the aquifer or by increasing the well diameter. Horizontal screen pipes, radial screen pipes, or horizontally dug wells are examples of well designs that increase injection flow capacity. These installations can increase the surface area interfacing the aquifer and the radius of influence of a well to achieve higher injection rates than with a traditional well.

Another option for increasing well injection flow may be the use of an infiltration pit dug near the bottom of a well. An infiltration pit could be used to increase to surface area of contact of the well to the aquifer in both saturated and unsaturated regions. This unused and unproven option may complicate well completion procedures and increase cost. Figure 4.13 illustrates the infiltration pit well concept.

The infiltration pit option presents different implementation challenges, depending on whether it is used in a confined or unconfined aquifer. The left half of Figure 4.13 shows a characteristic mound of injection in the unconfined case. One design consideration here is where to place the pump turbine unit. The water level in the well changes significantly during pumping and injection modes. To alleviate the problem of a "dry" pumping situation, an extension pipe that reaches toward the bottom of the completion is installed. Alternately, during turbine operation, it is desirable to allow free flow of water at the exit of the turbine. To accomplish this, a short pipe that dumps water into the air above the water level in the well is proposed. This allows maximum water velocity through the turbine, increasing generating efficiency. In the unconfined aquifer, the situation is more difficult. It is likely not possible to operate the turbine so that its exit water dumps into free air. Thus, the velocity of water through the turbine may not be ideal.

It is difficult to compare the performance of an infiltration pit well to the other completion options. The factors involved include pit size and dynamic flow patterns in the wells. Field testing of such well completions is needed to add to our knowledge about UPHES systems.

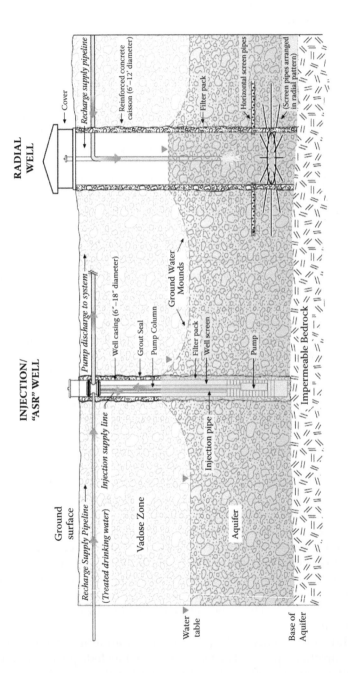

FIGURE 4.11

Direct injection radial unconfined aquifer well concept. (Courtesy of Topper, R., P.E. Barkmann, D.A. Bird, and M.A. Sares. 2004. *Artificial Recharge of Groundwater in Colorado: A Statewide Assessment.* Colorado Geological Survey, Department of Natural Resources, Denver, CO.)

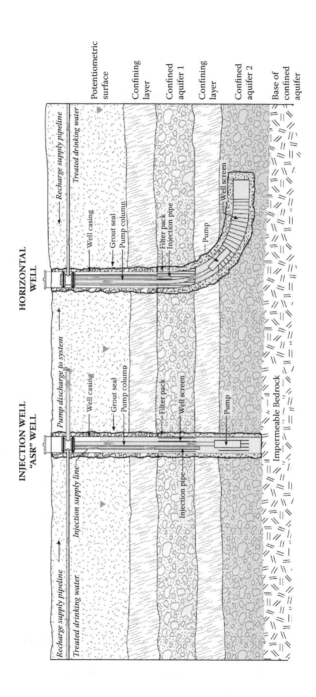

FIGURE 4.12
Direct injection horizontal confined aquifer well concept. (Courtesy of Topper, R., P.E. Barkmann, D.A. Bird, and M.A. Sares. 2004. *Artificial Recharge of Groundwater in Colorado: A Statewide Assessment.* Colorado Geological Survey, Department of Natural Resources, Denver, CO.)

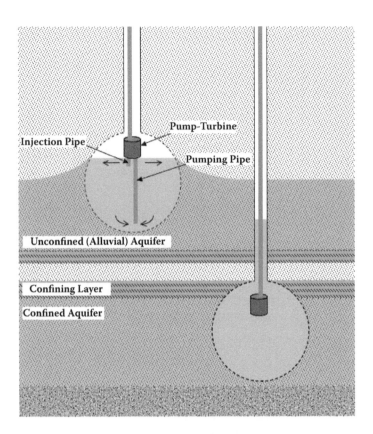

FIGURE 4.13
Infiltration pit wells in confined and unconfined aquifers.

Well modifications of the types outlined in this section are proposed for use in implementing aquifer UPHES systems. They include increased well radius, horizontal pipe completions, radial completions, horizontal "bending" well geometry, and infiltration pits. The best method of increasing well flow rates will depend on site-specific geology and aquifer characteristics, the availability of technology and tools to implement these advanced completions, and budget and power requirements.

Surface Reservoir

A surface water reservoir is needed to contain the water pumped up from the aquifer until it is used. Water pumped and held at the surface represents stored potential energy with respect to the aquifer. This energy can be converted back into electricity via a turbine generator or it can be partially allocated to another use such as irrigation. Surface ponds are not uncommon structures. Permitting, design, construction, and use of surface reservoirs are well understood and should pose no engineering challenges for

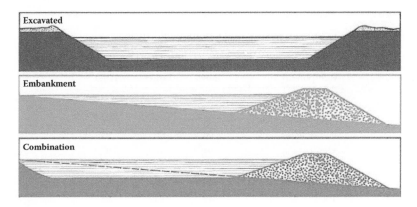

FIGURE 4.14
Major types of pond excavations. (From: http://www.dnr.state.oh.us/wildlife/Home/fishing/pond/construction/tabid/6218/Default.aspx. With permission.)

implementation in most cases. The cost of excavating and lining a new surface reservoir and the challenge of maintaining sufficient water quality are the major foreseeable concerns.

The most suitable type of reservoir depends on site characteristics such as topography, soil composition, and local regulations. The main types of reservoir (pond) designs are excavated, embankment, or a combination. Excavated ponds are more common on flat terrain; embankment ponds are commonly used with sloping terrain. Figure 4.14 illustrates these pond excavations.

In an aquifer UPHES system, the water level in the pond will rise and fall frequently. The magnitude of this change will depend on the volume and surface area of the reservoir with respect to the amount of water pumped or injected. If the pond water is to be used directly for crop irrigation, the volume of water must be sufficient to support both irrigation and aquifer injection volumes. As reservoir surface area increases, however, evaporation losses also increase. The reservoir volume and depth must be traded with the allowable water level change. While an aquifer UPHES consumes no net water during operation, a reservoir owner must have sufficient water rights to account for any losses due to evaporation or irrigation uses.

Excavation costs can range widely, depending on the soil type, size, and local labor rates and economics. Finally, a hydraulic interface that allows water to be pumped in and out of the reservoir is required for an aquifer UPHES. This will necessitate the installation of underground or aboveground water piping and valves interfacing the reservoir to the well.

System Efficiency

The efficiency of the operation of an aquifer UPHES system is an important measure of its feasibility. In this section, estimates of the efficiencies of the components and the resulting system efficiencies are provided. In previous

TABLE 4.1

Estimated System and Component Efficiencies

Component	Efficiency (%)		
	Low	Target	High
Pumping			
VFD pump drive	94	95	97
Power wires	96	98	99
Motor	94	96	97
Pump	60	70	75
Pipe friction	96	97	98
Total	49	61	68
Generating			
Pipe friction	96	97	98
Turbine	70	80	85
Generator	93	95	96
Rectifier	95	97	98
Inverter	94	96	97
Total	56	69	76
Round-trip efficiency	27	42	52

sections, discrete component efficiencies are introduced. These values are summarized in Table 4.1. The pump or turbine has the most impact on system efficiency. Electrical system components including the motor generator have relatively high efficiencies. One should note that the round trip efficiency is not the figure of merit for an aquifer UPHES. Rather, the turbine operation efficiency should be emphasized because, during pumping, energy that would otherwise be unused is used to pump water. Therefore, the pumping cycle can be viewed as "free" and the generating cycle viewed as the efficiency of merit for the system.

Aquifer Hydrogeology

The success of an aquifer UPHES installation depends on favorable hydrogeologic conditions. Aquifer hydrogeology is briefly discussed in this section, and typical values for important UPHES design parameters are introduced.

Aquifers fall into two major categories: unconfined and confined. Unconfined aquifers are also called water table or phreatic aquifers because their upper boundaries are the water tables. Usually, the most shallow aquifer at a given location is unconfined, with confined aquifers below. Unconfined and confined aquifers are separated by confining layers called aquitards or aquicludes—geologic formations of very low hydraulic conductivity. Unconfined aquifers generally receive recharge water from direct precipitation or from a body of surface water such as a river or lake.[15] Confined aquifers have water tables above their upper boundaries; thus a well dug into a confined aquifer may find pressurized water or even artesian flow to the surface.

The storage coefficient is an important characteristic that distinguishes confined and unconfined aquifers. Confined aquifers have very low storage coefficient values (generally less than 0.01 and as little as 10^{-5}). These values indicate that a confined aquifer stores water using the mechanisms of aquifer matrix expansion and the compressibility of water; both typically are quite small quantities. Unconfined aquifers have storage coefficients (specific yields) normally above 0.01, and they release water from storage by the mechanism of actually draining the pores of the aquifer, releasing relatively large amounts of water.

Both unconfined and confined aquifers are candidates for aquifer UPHES installation. Confined aquifers have the advantage of being much deeper (farther below the surface) than unconfined aquifers. However, the specific yields of confined aquifers are decidedly lower than those of unconfined aquifers. Alternately, while unconfined aquifers have high specific yield capacities, they are generally much shallower or closer to the surface. Here again, we see a design trade-off between a high head, low flow option and a low head, high flow option. Another important note is that water quality requirements are more stringent for unconfined aquifers. Table 4.2 gives a qualitative comparison of the two types of aquifers.

Considering the minimum requirements for an aquifer UPHES system, for a 200 foot (60.1 meter) thick aquifer, a transmissivity of 2.6 ft^2/min (40.3 cm^2/s) translates to a hydraulic conductivity of 0.013 ft/min (0.0066 cm/s). Table 4.3 summarizes the typical ranges of hydraulic conductivity and transmissivity values for different geologic materials. Based on the ranges in the table, unconsolidated gravel and sand, sedimentary limestone, dolomite,

TABLE 4.2

Qualitative Comparison of Aquifer Types

	Unconfined Aquifer	**Confined Aquifer**
Hydraulic conductivity	Medium to high	Low to medium
Storage coefficient	Medium to high	Low
Transmissivity	Medium to high	Low to medium
Depth to water	Low to medium	Low to high
Specific yield	High	Low to medium
Advantages	Existing irrigation wells, high flow yield	Very high head potential, water quality specifications easily met
Disadvantages	Stringent water quality specificationss, water rights difficult to obtain, depth to water typically shallow	Low flow yield, advanced completion more difficult to use

TABLE 4.3

Typical Ranges of Hydraulic Conductivity and Transmissivity in Aquifer Materials

Material	Conductivity k [cm/s]		Conductivity k [ft/min]		Transmissivity [cm²/s], 61-m depth		Transmissivity [ft²/min], 200 ft depth	
	Min	Max	Min	Max	Min	Max	Min	Max
Unconsolidated								
Gravel	1.0E-01	1.0E+01	2.0E-01	2.0E+01	610	60960	39	3936
Sand	1.0E-04	1.0E+00	2.0E-04	2.0E+00	0.610	6096	0.039	394
Silt	1.0E-07	1.0E-03	2.0E-07	2.0E-03	0.001	6.096	0.000039	0.394
Clay and glacial till	1.0E-11	1.0E-06	2.0E-11	2.0E-06	0.0000000610	0.006	0.000000004	0.000394
Sedimentary rock								
Sandstone	1.0E-08	1.0E-03	2.0E-08	2.0E-03	0.000061	6.096	0.000004	0.394
Limestone, dolomite	1.0E-07	1.0E-01	2.0E-07	2.0E-01	0.001	610	0.000039	39
Karst limestone	1.0E-04	1.0E+00	2.0E-04	2.0E+00	0.610	6096	0.039	394
Shale	1.0E-11	1.0E-06	2.0E-11	2.0E-06	0.000000061	0.006	0.0000000039	0.000394
Crystalline rock								
Basalt	1.0E-09	1.0E-05	2.0E-09	2.0E-05	0.000006	0.061	0.0000003936	0.004
Fractured basalt	1.0E-05	1.0E+00	2.0E-05	2.0E+00	0.061	6096	0.004	394
Dense crystalline rock	1.0E-12	1.0E-08	2.0E-12	2.0E-08	0.0000000061	0.000061	0.0000000004	0.000004
Fractured crystalline rock	1.0E-06	1.0E-02	2.0E-06	2.0E-02	0.006	60.960	0.000394	3.936

Source: Becker, M.F. et al. 1999. Groundwater quality in the Central High Plains Aquifer of Colorado, Kansas, New Mexico, Oklahoma, and Texas. U.S. Geologic Survey, WRIR 02-4112.

Karst limestone, and crystalline fractured basalt aquifer geologies are candidates for aquifer UPHES. As an example, hydraulic conductivities in the high plains Ogallala aquifer in the central United States generally lie in the range of 25 to 100 feet per day (0.017 ft/min to 0.07 ft/min or 0.0086 cm/s to 0.036 cm/s) with an average estimated at 51 feet daily (0.035 ft/min or 0.018 cm/s).[16] In addition, the maximum thickness of this aquifer can exceed 700 feet (213.4 meters). Using these ranges, the transmissivities for several cases are calculated and shown in Table 4.4. The table selects minimum, maximum, and median values within the above ranges for hydraulic conductivity and thickness, and an average expected transmissivity is calculated to be 17.4 ft^2/min (269.4 cm^2/s). The depth to water and flow yield of a well are important parameters to lead the search for a suitable aquifer UPHES site.

Legal Considerations

Before building a UPHES system, a permit is normally required from the state in which the system will be installed. In addition, several state and federal regulations relating to water usage, contamination, quality, and land use must be followed. This section briefly discusses some of the major regulations[17] to consider when planning a UPHES installation.

Tributary ground water is generally considered "water of every natural stream" and is subject to appropriation and regulation. The basis for this classification is the hydrological connection of this ground water to surface water. Legally, tributary ground water is generally treated the same as

TABLE 4.4

Transmissivity Averaging Calculations for Ogallala Aquifer

Hydraulic Conductivity [ft/day]	Hydraulic Conductivity [ft/min]	Thickness [ft]	T [ft^2/min]
25	0.017	100	1.74
25	0.017	400	6.94
25	0.017	700	12.15
50	0.035	100	3.47
50	0.035	400	13.89
50	0.035	700	24.31
75	0.052	100	5.21
75	0.052	400	20.83
75	0.052	700	36.46
100	0.069	100	6.94
100	0.069	400	27.78
100	0.069	700	48.61
		Average	17.36

surface waters (e.g., rivers and streams). The provisions of the Water Right Determination and Administration Act of 1969, as modified since original enactment, govern the use of natural stream waters including tributary ground water.

Because tributary aquifer ground water is contained in an aquifer that is directly connected to the local stream system, generally the water table in such an aquifer is relatively shallow. On the other hand, deep aquifer ground water is not so directly connected to the surface stream system (i.e., non-tributary ground water is more likely to be deep aquifer ground water). A site using non-tributary ground water may better meet the needs (i.e., head requirements) for an aquifer UPHES system. Other advantages are associated with the non-tributary regulatory scheme, for example, the manner in which water rights are allocated and the accounting mechanisms for water use. Because an aquifer UPHES will put water to new and different use, a change of water right may need to be undertaken for both tributary and non-tributary ground water. An application for a change of water right may have to be pursued through the relevant water court.

Another question that arises is whether a storage right must be obtained for the surface impoundment. Because an aquifer UPHES will utilize water by storing it in a surface impoundment for later use rather than putting it to direct use (such as for irrigation), the installation may require a storage right.

If well equipment must be modified to accommodate an aquifer UPHES system, a new well permit may be necessary. Further, well construction requirements pursuant to federal laws and regulations may apply to re-injection of water into underground sources. Underground injection permitting will be required for an aquifer UPHES system.

The U.S. Environmental Protection Agency (EPA) regulates water re-injection back into aquifers and water sources and also regulates and enforces water quality issues. The drainage of water from a surface impoundment down to an underground source or aquifer is subject to regulations regarding water quality. Class V injection well requirements under the federal Safe Drinking Water Act (SDWA) apply. Protection of other water rights, including the quality of water of that right, is required.

Because of ground water contamination occurrences in the 1960 and 1970s resulting from underground injection, Congress passed the SDWA in 1974. Part C of the act required EPA to establish a system of regulations for injection activities (42 USC §§300h et seq.). The regulations establish minimum requirements for controlling all injection activities and provide mechanisms for implementation and authorization of enforcement authority along with protection for underground sources of drinking water.

Historically, the SDWA has applied to water returned to an underground source through aquifer recharge or aquifer storage recovery (ASR) wells. However, based on the definition of "well" and the lack of applicable exclusions, it appears that this act applies to aquifer UPHES systems contemplated here as Class V wells. The Underground Injection Control (UIC) program defines a well as any bored, drilled, or driven shaft or a dug hole whose depth is greater than the largest surface dimension used to discharge fluids underground. To comply, the owner or operator of a Class V well is required to submit basic inventory information and operate the well such that drinking water is not endangered. Note that because ASR and aquifer recharge wells are authorized by rule, they do not require permits unless required to do so by the Underground Injection Control (UIC) Program Director under 40 CFR §144.25.

Further regulations, rules, and permitting specifications may be mandated by the state in which an aquifer UPHES project will reside. These regulations and activities required to meet them must be evaluated and understood. Some generalities can be made about site preferences for such systems. Designated basins will probably not be advantageous sites for a number of reasons. For example, the depths of wells associated with designated basins are typically too shallow for the necessary head; these are typically over-appropriated water sources. The permitting process is more complex. Between tributary and non-tributary sources, non-tributary types appear more advantageous because of the manner in which water rights are allocated and the accounting mechanisms for water use. In addition, wells for non-tributary water sources will usually be deeper.

Economics

The cost of an aquifer UPHES system depends on many factors, many of which are site-specific. The amount of well modification required, the presence of an existing surface reservoir, and the possibility of using existing irrigation machinery may significantly reduce the total system cost. Site characteristics such as transmissivity and depth to water exert important effects on the cost of a system. Designers must strive to locate aquifer UPHES systems in areas where the beneficial pameters are maximized.

Compiling a levelized cost estimation for an aquifer UPHES system may be instructive, but the result depends heavily on the assumptions made. In this section, an attempt is made to suggest the expected levelized cost of energy associated with such energy storage systems.

Levelized cost is defined as the cost per unit energy of the installation, averaged over its lifetime. Cost ranges for an aquifer UPHES are estimated below for a system sized to provide up to 300 kWh of energy per cycle. It should be noted that this is an energy storage system, so rather than producing energy, it consumes a small amount (due to efficiency losses). Thus, the levelized cost calculated here is applicable to an energy storage system only—one that is not coupled to a generating source and does not produce electricity. The following is a list of assumptions made for the purpose of levelized cost estimation:

System rated power = 50 kW
System rated energy = 300 kWh per cycle
Number of cycles per day = 1
Number of days operating per year = 150
Operating lifetime of system = 35 years
System lifetime capital and operating cost = $300,000
System round-trip efficiency = 50%
Photovoltaic solar system levelized cost = $0.03 per kWh

The levelized cost of the stand-alone system is found by summing the total energy (stored) over the lifetime of the system and then dividing the total costs by this energy result:

$$(300 \text{ kWh}) \cdot (150 \text{ days}) = 45,000 \text{ kWh per year}$$

$$(45000 \text{ kWh/year}) \cdot (35 \text{ years}) = 1,575,000 \text{ lifetime kWh}$$

$$(\$300,000)/(1,575,000 \text{ kWh}) = \$0.19 \text{ per kWh stored}$$

This result, a levelized cost of 19¢ per kWh, is higher than the cost of energy from most generating sources. However, this cost cannot be directly compared to generating costs because this system does not generate. The value of the storage system lies in the ability to capture variable or low cost energy and deploy it as needed.

Future Prospects

Underground pumped hydroelectric energy storage is a feasible means of storing energy that has not been comprehensively analyzed. Further, no UPHES system has ever been built. Research and analysis are desperately needed to examine possible UPHES installations, both large and small. Over the past three decades, enabling technologies needed for UPHES have grown, matured, and become more efficient. These include excavation techniques, high-head hydro machinery, and geologic analysis techniques. Technologies developed to extract oil and coal deep below the surface provide starting points for building UPHES systems. The current growth of variable renewable energy that could greatly benefit from economical large and small scale energy storage further drives the need for rekindled UPHES interest.

References

1. Chiu, H.H., Saleem, Z.A., Ahluwalia, R.K. et al. 1978. Mechanical energy storage systems: compressed air and underground pumped hydro. 16th AIAA Aerospace Sciences Meeting.
2. Tam, S.W., Blomquist, C.A., and Kartsounas, G.T. 1979. Underground pumped hydro storage: an overview. *Energy Sources* 4, 329.
3. Allen, R.D., Doherty, T.J., and Kannberg, L.D. 1984. Underground pumped hydroelectric storage. Pacific Northwest National Laboratory, Richland, WA.
4. Ramer, J.L. August 4, 1981. U.S. Patent 4,282,444.
5. Chen, H.H., and Berman, I.A. September 1982. Commonwealth Edison Company's underground pumped hydro project. AIAA/EPRI International Conference on Underground Pumped Hydro and Compressed Air Energy Storage, San Francisco.
6. Uddin, N. 2003. Preliminary design of an underground reservoir for pumped storage. *Geotechnical and Geological Engineering* 21, 331.
7. Martin, G. 2007. Aquifer Underground Pumped Hydroelectric Energy Storage. M.S. Thesis, University of Colorado at Boulder.
8. Charbeneau, R.J. 2000. *Groundwater Hydraulics and Pollutant Transport*. Prentice-Hall, New York.
9. Williams, A.A. 1994. Turbine performance of centrifugal pumps: comparison of prediction methods. *Journal of Power and Energy* 208.
10. Gordon, J.L. 2001. Hydraulic turbine efficiency. *Canadian Journal of Civil Engineering* 28, 238.
11. Chan, T.F. 1993. Capacitance requirements of self-excited induction generators. *IEEE Transactions on Energy Conversion* 8, 304.
12. Mohamed, A., Masoum, S., Dehbonei, H. et al. 2002. Theoretical and experimental analyses of photovoltaic systems with voltage- and current-based maximum power point tracking. *IEEE Transactions on Energy Conversion* 17, 514.
13. City of Tampa (FL) Water Department. *2003 Annual Report*, p. 28.
14. Topper, R., Barkmann, P.E., Bird, D.A., and Sares, M.A. 2004. *Artificial Recharge of Groundwater in Colorado: A Statewide Assessment*. Colorado Geological Survey, Department of Natural Resources, Denver.
15. Emery, P.A. *Hydrogeology of the San Luis Valley, Colorado: An Overview and a Look at the Future*. http://www.nps.gov/grsa/naturescience/upload/Trip2023.pdf
16. Becker, M.F. et al. 1999. Groundwater quality in the Central High Plains Aquifer of Colorado, Kansas, New Mexico, Oklahoma, and Texas. U.S. Geologic Survey, WRIR 02-4112.
17. Ginocchio, A., Lent-Parker, M., and Sewalk, S. 2007. *Review of the Legal and Regulatory Requirements Applicable to a Small-Scale Hydroenergy Storage System in an Agricultural Setting*. Energy and Enrivonmental Security Initiative, University of Colorado School of Law.

5

Compressed Air Energy Storage

Samir Succar

CONTENTS

Background

Compressed air energy storage (CAES) is a low cost technology for storing large quantities of electrical energy in the form of high-pressure air. It is one of the few energy storage technologies suitable for long duration (tens of hours), utility scale (hundreds to thousands of megawatts) applications. Several other energy storage technologies such as flywheels and ultracapacitors can provide short duration services related to power quality and stabilization but are not cost effective options for load shifting and wind generation support [1,2].

The two principal technologies capable of delivering several hours of output at a plant-level output scale at attractive system costs are CAES and pumped hydroelectric storage (PHES) [3–8]. Although some emerging battery technologies may provide wind-balancing services as well, typical system capacities and storage sizes are an order of magnitude smaller than CAES and PHES systems (~10 MW, <10 hours) with significantly higher capital costs.

PHES does not require fuel combustion and has more field implementation than CAES, but is economically viable only at sites where reservoirs at differential elevations are available or can be constructed at manageable cost. Furthermore, the environmental impacts of large-scale PHES facilities are becoming more complex, especially where preexisting reservoirs are not available and sites with large, natural reservoirs at large differential elevations where environmentally benign, inexpensive PHES facilities can be built are increasingly rare.

In contrast, CAES can use a broad range of reservoirs for air storage and has a more modest surface footprint, giving it greater siting flexibility relative to PHES. High pressure air can be stored in surface piping, but for large-scale applications, developing storage reservoirs in underground geologic formations such as solution-mined salt, saline aquifer, abandoned mine, and mined hard rock are typically more cost effective. The widespread availability of geologies suitable for CAES in the continental United States suggests that this technology faces far fewer siting constraints than PHES—especially important for deploying CAES for wind balancing.

One of the main applications for CAES is for the storage of wind energy during times of transmission curtailment and generation onto the grid during shortfalls in wind output. Wind balancing requires large-scale, long duration storage, fast output response times, and siting availability in wind-rich regions. Prior studies indicate that suitable CAES geologies are widely available in the wind-rich Great Plains of the U.S. Furthermore, CAES is able to ramp output quickly and operate efficiently under partial load conditions, making it suitable for balancing fluctuations in wind energy output. Finally, the low greenhouse gas (GHG) emissions rate of CAES makes it a good candidate for balancing wind in a carbon constrained world. Several prior studies have analyzed the costs and emissions of hybrid wind–CAES systems [9–18].

Among the geologic options for air storage, porous rock formations offer the best availability and potentially lowest cost. Moreover, geographical distributions of aquifers and good wind resources are strongly correlated in the U.S. Therefore, the potential for CAES to play a major role in balancing wind output and producing low GHG emitting power will depend to a large degree on the availability of aquifer structures suitable for CAES.

Evolving Motivations for Bulk Energy Storage

CAES emerged in the 1970s as a promising peak shaving option [19]. High oil prices together with an expanding nuclear power industry sparked an interest in energy storage technologies such as CAES to be used in load-following applications. The high price of peak power and the perceived potential for inexpensive baseload nuclear power made attractive the option of storing inexpensive off-peak electricity and selling this electricity during peak demand periods [20,21].

These conditions initially fueled a strong interest in CAES by many utilities, but as the nuclear power industry lost momentum and oil prices retreated from their peaks, the market conditions for CAES began to change. During the 1980s, gas turbines and combined cycle generation emerged as the leading low cost options for peaking and load-following markets. The new options along with overbuilt generating capacity on the grid and the perception that domestic natural gas supplies were abundant led to erosion of market interest in energy storage.

Recent trends in wind power development have fostered new interest in energy storage, not as a way to convert baseload power into peak power, but as a way mitigate the variability of wind energy [16,18]. Global wind power capacity has grown rapidly in recent years from 4.8 GW in 1995 to 121 GW by the end of 2008 (Figure 5.1). The variability of wind output requires standby reserve capacity to ensure output during peak demand. Gas turbines can respond quickly to shortfalls in wind output and thus gas fired spinning reserve units are good candidates for dispatch to meet the challenge of balancing this growing wind segment of the power mix.

Energy storage represents an alternative wind balancing strategy, and the low fuel consumption of CAES makes it especially attractive during times of high and/or volatile gas prices. Although wind balancing has long been acknowledged as a potential application for bulk energy storage [22], only recently has wind penetration reached levels that require additional balancing measures for maintaining system stability [23].

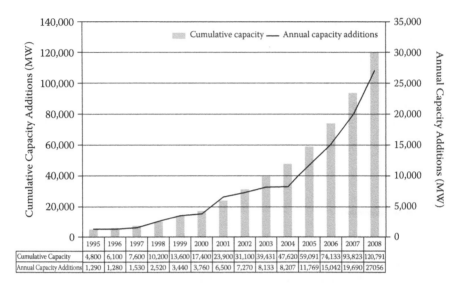

	1995	1996	1997	1998	1999	2000	2001	2002	2003	2004	2005	2006	2007	2008
Cumulative Capacity	4,800	6,100	7,600	10,200	13,600	17,400	23,900	31,100	39,431	47,620	59,091	74,133	93,823	120,791
Annual Capacity Additions	1,290	1,280	1,530	2,520	3,440	3,760	6,500	7,270	8,133	8,207	11,769	15,042	19,690	27056

FIGURE 5.1
Global wind capacity, 1995 to 2008. (Global Wind Energy Council, Brussels, 2009.).

Recent studies show that bulk storage can reduce the integration costs for wind energy even at relatively low penetration levels [24]. The use of storage for balancing wind and serving other grid management applications will be especially valuable where the supply of flexible generating capacity (e.g., hydroelectric) is limited [18,25]. The continued increase of wind penetration on the grid and the need to reduce GHG emissions may create an incentive to use storage systems directly coupled with wind to produce baseload power rather than as independent entities to provide grid support (see below). Further, because the fuel consumption of CAES is less than half that of a simple cycle gas turbine, it provide a hedge against natural gas price volatility [26].

A further reason for considering wind farms coupled to CAES storage (wind/CAES) stems from the fact that most high quality onshore wind resources are often remote from load centers. The exploitable onshore wind potential in classes 4 and above in North America is huge—more than 12 times total electricity generation in 2004 [27,28]. However, wind resources in the United States are concentrated in the sparsely populated Great Plains and Midwest states that account for over half of the exploitable wind generation potential in class 4+ [29]. Bringing cost-effective electricity from the Great Plains to major urban electricity demand centers requires that it be transmitted via baseloaded GW-scale high-voltage transmission lines. CAES systems coupled to multiGW-scale wind farms could provide such baseload power.

Because the incremental capital cost for increasing CAES storage volume capacity is relatively low, CAES is well suited for providing long-duration storage (>80 hours) needed to produce baseload power. Although seasonal storage of wind is also possible, it would require much larger storage volumes [30]. While typical capacity factors for wind farms are approximately 30 to 40% [31], wind/CAES systems can achieve capacity factors of 80 to 90% typical of baseload plants [10]. Therefore, the coupling of wind to energy storage enhances utilization of both existing transmission lines and dedicated new lines for wind. This can alleviate transmission bottlenecks and minimize the needs for transmission additions and upgrades. The cited report indicates that removal of bulk storage (pumped hydroelectric storage in this case) increases integration costs for wind by approximately 50% for a wind penetration level of 10%. Also, doubling of storage capacity lowered integration cost by ~$1.34/MWh in the 20% penetration case.

The Greenblatt (2005) estimate is based on the assumption that various land use constraints limit the technical potential for wind to what can be produced on 50% of the land on which class 4+ wind resources are available. The technical wind power potential at the global level is also huge. Considering only class 4+ winds exploited on 50% of the land on which these resources are available, as in the North American case, Greenblatt (2007) estimated that the global technical wind energy potential is 185,000 TWh/year on land and 49,400 TWh/year offshore. For comparison, the global electricity generation rate in 2004 was 17,400 TWh/year.

Capacity factor in this case is on the basis of a constant demand level. The rated capacity of the wind park will be "oversized" relative to this demand level and the CAES turbo expander capacity matched to it such that excess wind can be stored to balance subsequent shortfalls. While it is possible to produce constant output (i.e., 100% capacity factor) from a wind/CAES plant, significantly larger storage would be required.

Where transmission capacity is limited, it will be advantageous to site the storage reservoir and wind turbine array as closely as possible to exploit the benefits described above. If this is not possible, there is no need to co-locate the storage system and wind array. Independently siting these components would allow added flexibility for simultaneously matching facilities to the ideal wind resource, storage reservoir geology, and the required natural gas supplies.

System Operation

CAES systems operate in much the same way as conventional gas turbines except that compression and expansion operations occur independently and at different times (Figure 5.2). Because compression energy is supplied separately, the full output of the turbine can be used to generate electricity during expansion, whereas conventional gas turbines typically use

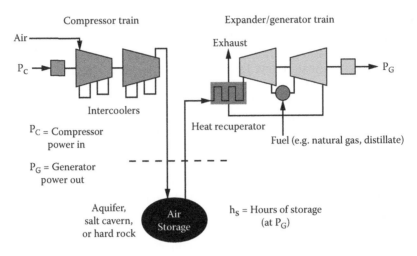

FIGURE 5.2
CAES system configuration.

approximately two thirds of the output power from the expansion stage to run compressors.

During the compression (storage) mode, electricity is used to run a chain of compressors that inject air into an uninsulated storage reservoir, thus storing the air under high pressure and at the temperature of the surrounding formation. The compression chain makes use of intercoolers and an aftercooler to reduce the temperature of the injected air, thereby enhancing compression efficiency, reducing the storage volume requirement, and minimizing thermal stress on the storage volume walls.

Despite the loss of heat via compression chain intercoolers, the theoretical efficiency for storage at formation temperatures in a system with a large number of compressor stages and intercooling can approach that for a system with adiabatic compression and air storage in an insulated cavern. Furthermore, despite the higher input energy required per unit mass due to cooling needs, overall fuel consumption is still dramatically lower since the net output of CAES is three times that of a conventional turbine [32].

During expansion (generation) operation, air is withdrawn from storage and fuel (typically natural gas) is combusted in the pressurized air. The combustion products are then expanded (typically in two stages), thus re-generating electricity. Fuel is combusted during generation for capacity, efficiency, and operational considerations. Expanding air at the wall temperature of the reservoir would necessitate much higher air flow to achieve the same turbine output, thus increasing the compressor energy input requirements to the extent that the charging energy ratio would be reduced by roughly a factor of four [33]. Furthermore, in the absence of fuel combustion, the low temperatures at the turbine outlet would pose a significant icing risk for the blades because of the large airflow through the turbine, despite the small specific moisture content for air at high pressure. Another possibility is that the turbine materials and seals may become brittle during low temperature operation.

Adiabatic CAES designs capture the heat of compression in thermal energy storage units. For example, assuming air recovered from storage at 20°C, adiabatic expansion, and a 45× compression ratio, T = 174°C at the turbine exhaust.

Suitable Geologies for CAES

Geologies suitable for CAES storage reservoirs can be classified as salt, hard rock, and porous rock. The total areas that have one or more of these geologies account for a significant fraction of the continental United States (Figure 5.3). Studies indicate that over 75% of the United States has geologic conditions that are potentially favorable for underground air storage [34,35].

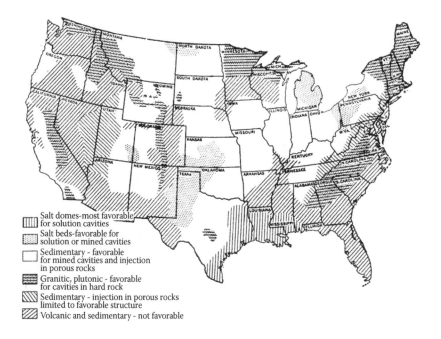

FIGURE 5.3

Areas classified for subsurface storage of fluids. (From National Petroleum Council Report of Committee on Underground Storage for Petroleum, April 22, 1952; updated October 1962 by C.T. Brandt. Bartlesville, OK; see Reference 19.)

However, those studies carried out only macro scale analyses that did not evaluate areas according to the detailed characteristics necessary to fully estimate their suitabilities for CAES. While the large fractions of land possessing favorable geologies appear encouraging, broad surveys such as the data presented in Figure 5.3 can serve only as templates for identifying candidate areas for further inquiry. Detailed regional and site-specific data will be necessary to determine the true geologic resource base for CAES installation.

Salt Geology

The two CAES plants currently operating use solution-mined cavities in salt domes as their storage reservoirs (see Figure 5.4 and the section covering existing and proposed CAES plants). In many ways, such formations are the most straightforward to develop and operate. Solution mining techniques can provide reliable, low cost routes for developing storage volumes of the needed size (typically at a storage capital cost of ~$2/kWh of output from storage) if an adequate supply of fresh water and efficient disposal of

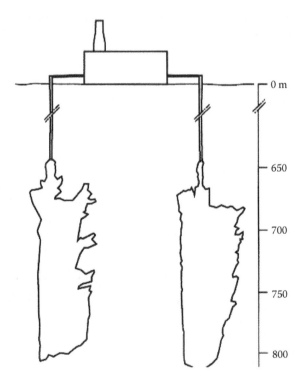

FIGURE 5.4
Structure of Huntorf CAES plant salt dome storage with caverns and plant on same scale. (From F. Crotogino, K. U. Mohmeyer, and R. Scharf, Huntorf CAES: More than 20 Years of Successful Operation, Solution Mining Research Institute Meeting, Orlando, FL, 2001.)

the resulting brine are available [1,2]. Furthermore, due to the elasto-plastic properties of salt, storage reservoirs solution-mined from salt pose minimal risk of air leakage [34,37]. However, brine disposal, cavern "rat holes," creep, and turbine contamination remain potential challenges [32].

Large bedded salt deposits are available in areas of the Central, North Central and North East United States and domal formations can be found in the Gulf Coast basin [38]. Although both formation types can be used for CAES, salt beds are often more challenging to develop when large storage volumes are required. Salt beds tend to be much thinner and often contain comparatively higher concentrations of impurities that present significant challenges with respect to structural stability [38]. Caverns mined from salt domes can be tall and narrow with minimal roof spans as is the case at both the Huntorf (see Figure 5.4) and McIntosh CAES facilities. The thinner salt beds cannot support long aspect ratio designs because the air pressure must support much larger roof spans. In addition, impurities may further compromise the structural integrity of a cavern and further complicate the development a large capacity storage system.

Although the locations of domal formations in the United States are not well correlated with high quality wind resources (see Figure 5.7), there are some indications the prospects may be more favorable in Europe (see Figure 5.5).

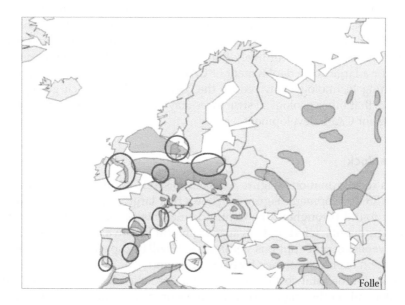

FIGURE 5.5
Coincidence of high wind potential and salt domes in Europe. Circles indicate areas investigated for CAES development. (From B. Calaminus, Innovative adiabatic compressed air energy storage system of EnBW in Lower Saxony, Second International Renewable Energy Storage Conference, Bonn, 2007.)

Hard Rock

Although hard rock is an option for CAES design, the cost of mining a new reservoir is often relatively high (typically $30/kWh produced). However, in some cases existing mines may be used, in which case the cost will typically be about $10/kWh produced [1,40,41] as is the case for the proposed Norton CAES plant that plans to use an idle limestone mine.

Detailed methodologies have been developed for assessing rock stability, leakage, and energy loss in rock-based CAES systems including concrete-lined tunnels [45–47]. Several such systems have been proposed [48] and known field tests include two recent programs in Japan: a 2 MW test system using a concrete-lined tunnel in the former Sunagaawa coal mine and a hydraulic confinement test performed in a tunnel in the former Kamioka mine [1].

In addition, a test facility was developed and evaluated by the Electric Power Research Institute (EPRI) and Luxembourg's Societé Electrique de l'Our SA utility using an excavated hard rock cavern with water compensation [49]. The site was used to determine the feasibility of such a system for CAES operation and characterize and model water flow instabilities resulting from the release of dissolved air in the upper portion of the water shaft (i.e., the "champagne effect").

Hard rock geologies suitable for CAES are widely available in the continental United States and overlap well with high-quality wind resources [82]. However, because development costs are high relative to other geologies (especially given the limited availability of preexisting caverns and abandoned mines [37]), it is unlikely that this option will be the first one pursued for a large-scale deployment of CAES. Although future developments in mining technology may reduce the costs of utilizing such geologies, it appears that other geological structures may offer the best near-term opportunities for CAES development.

Porous Rock

Porous rock formations (Figure 5.6) such as saline aquifers are also suitable for CAES development. Figure 5.7 shows that large, homogeneous aquifers can be found throughout the Central United States. Because this area also has high quality wind and because of the limited availability and/or cost effectiveness of other options, aquifer CAES will be especially relevant to the discussion of energy storage for balancing wind.

Although the total cost of developing a porous rock formation for CAES will depend on the characteristics of the storage stratum (e.g., thinner, less permeable structures will require more wells and therefore higher development costs), it appears that this type of geology is often the lowest cost option. CAES estimates indicate that total development costs are in the range $2 to $6 million/Bcf of total volume (working gas and base gas)—similar to

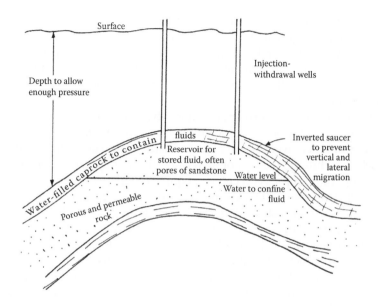

FIGURE 5.6
Porous rock CAES storage volume with significant wind power generation. (From D. L. Katz and E. R. Lady. 1976. *Compressed Air Storage for Electric Power Generation*. Ulrich, Ann Arbor, MI. With permission.)

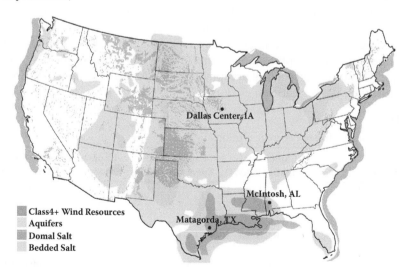

FIGURE 5.7
Comparison of areas of high quality wind resources and geology compatible with CAES (suitable for mined rock caverns eliminated due to high estimated cost of developing them for CAES. Locations of existing McIntosh CAES plant, recently announced Dallas Center wind/CAES system, and proposed Matagorda plant are indicated. (From D. L. Katz and E. R. Lady. 1976. *Compressed Air Storage for Electric Power Generation*. Ulrich, Ann Arbor, MI. With permission.)

development cost estimates for natural gas storage in similar formations [50]. This implies a capital cost of $2 to $7/kWh of storage capacity, depending on the site characteristics and assuming a 5:1 base gas-to-working gas volume ratio [51]. These costs are somewhat lower than those estimated for salt cavern storage ($6 to $10/kWh of storage capacity)—the next most economical option (see Table 5.1).

Aquifer CAES has the further advantage that the cost of incremental additions to storage capacity is significantly lower than for alternative geologies. Assuming sufficient wells are in place to ensure adequate air flow to the surface turbo machinery, the cost of increasing the storage capacity of an aquifer is simply the compression energy required to increase the volume of the bubble [52]. This cost (~$0.11/kWh) is an order of magnitude lower than the equivalent marginal costs of solution mining salt and more than two orders smaller than excavating additional cavern volume from hard rock [1].

Despite these low development costs and apparent widespread availability, extensive characterization of candidate formations is required to determine project feasibility. Detailed measurements of permeability, porosity, and structure are required to determine a formation's suitability for storage operation [52]. Prior industrial experience with natural gas storage will be valuable as many of the methodologies used to characterize formations and develop projects are directly applicable to CAES development in an aquifer [53]. The industry's extensive experience with natural gas storage provides a theoretical and practical framework for describing underground storage media and assessing candidate sites for seasonal storage of natural gas. Storage capacity assessments for CO_2 storage may be helpful as well, although minimum depth required for CO_2 to become supercritical (~800 m) is typically at the high end of acceptable limits for CAES due to high pressure turbine inlet limitations.

TABLE 5.1

Capital Costs for Energy Storage Options

Technology	Capital Cost: Capacity ($/kW)	Capital Cost: Energy ($/kWh)	Hours of Storage	Total Capital Cost ($/kW)
CAES (300 MW)	580	1.75	40	650
Pumped hydroelectric (1,000 MW)	600	37.5	10	975
Sodium sulfur battery (10 MW)	1720 to 1860	180 to 210	6 to 9	3100 to 3400
Vanadium redox battery (10 MW)	2410 to 2550	240 to 340	5 to 8	4300 to 4500

Sources: Electric Power Research Institute and U.S. Department of Energy. 2003. *Handbook of Energy Storage for Transmission and Distribution Applications.* Palo Alto, CA and Washington; Electric Power Research Institute and U.S. Department of Energy. 2004. *Energy Storage for Grid-Connected Wind Generation Applications.* Palo Alto, CA and Washington; Electric Power Research Institute. 2005. *Wind Power Integration: Energy Storage for Firming and Shaping,* Palo Alto, CA.

While natural gas storage provides an important departure point for a discussion of CAES, several important differences must be considered, including differences in physical properties of the working fluid (e.g., viscosity, gas deviation factor) and new oxidation and corrosion mechanisms resulting from the introduction of oxygen into a formation. Also, a CAES system used for voltage regulation or backing wind power will likely switch between compression and generation several times a day. In contrast, most natural gas storage facilities are often only cycled once over the course of a year to meet seasonal demand fluctuations for natural gas. These are important differences that must be considered, but a wide range of advanced design concepts and mitigation techniques can be employed to address such requirements.

Although no commercial systems have been built to date, several successful field tests have demonstrated the operational feasibility of using an aquifer for compressed air storage applications. A 25 MW porous rock-based CAES test facility operated for several years in Sesta, Italy. Although the tests were successful, a geologic event disturbed the site and led to closure of the facility [1]. In addition, EPRI and the U.S. Department of Energy conducted tests on porous sandstone formations in Pittsfield, Illinois, to determine their feasibility for CAES. Testing for the first commercial CAES plant with a porous rock reservoir was scheduled to begin in Dallas Center, Iowa, in 2010.

In addition to using saline aquifers for CAES, it is also possible to use depleted oil and gas reservoirs that are fundamentally aquifers. Since the bulk of natural gas storage experience is in depleted fields, many issues related to residual hydrocarbons have been extensively studied; however, the injection of oxygen would present challenges not encountered when storing natural gas. In particular, the presence of residual hydrocarbons may introduce the risk of flammability and in situ combustion upon the introduction of high pressure air.

The flammability of the natural gas–air mixture may be another concern for CAES operation, but displacement of natural gas away from the active bubble area can mitigate this risk considerably. In some cases, nitrogen injection may be desirable to further minimize air–natural gas mixing. Previous studies indicate that these methods adequately address the challenge of using depleted natural gas fields for CAES and that these structures can provide suitable air storage media [53].

Existing and Proposed CAES Plants

Huntorf

The Huntorf CAES plant near Bremen, Germany, the world's first such facility, was completed in 1978 (see Figure 5.8 and Figure 5.9). The 290 MW plant was designed and built by ABB (formerly BBC) to provide black-start services

FIGURE 5.8
Aerial view of Huntorf CAES plant near Bremen, Germany. (From F. Crotogino, K. U. Mohmeyer, and R. Scharf, Huntorf CAES: More Than 20 Years of Successful Operation, Solution Mining Research Institute Meeting, Orlando, FL, 2001.)

FIGURE 5.9
Machine hall of Huntorf CAES plant. (From A. J. Karalis, E. J. Sosnowicz, and Z. S. Stys, Air storage requirements for a 220 M We CAES plant as a function of turbo machinery selection and operation, *IEEE Transactions on Power Apparatus and Systems*, 104: 803, 1985.)

to nuclear units near the North Sea and to furnish inexpensive peak power. [Note: Black start is the ability of a plant to start up during a complete grid outage.] Because nuclear power stations require some power to resume operation, the Huntorf plant was built in part to provide black-start power. It has operated successfully for over three decades, primarily as a peak shaving unit and to supplement other (hydroelectric) storage facilities on the system to fill the generation gap left by slow-responding medium-load coal plants. Availability and starting reliability for this unit are reported as 90 and 99%, respectively.

Because Huntorf was designed for peaking and black-start applications, it was initially designed with a storage volume capable of 2 hours of rated output. The plant has since been modified to provide up to 3 hours of storage and has been used increasingly to help balance the rapidly growing wind output from North Germany [36,54]. The underground portion of the plant consists of two salt caverns (310,000 m^3 total) designed to operate between 48 and 66 bar. The air from the salt caverns was found to cause oxidation upstream of the gas turbine during the first year of operation, leading to the installation of fiberglass reinforced plastic (FRP) tubing. Because the turbine expanders are sensitive to salt in the combustion air, special measures were taken to ensure acceptable conditions were met at the turbine inlet as well [36].

The compression and expansion sections draw 108 and 417 kg/s of air, respectively, and each is comprised of two stages. The first turbine stage expands air from 46 bar to 11 bar. Because gas turbine technology was not compatible with this pressure range, steam turbine technology was chosen for the high-pressure (hp) expansion stage. Due to the increase in heat transfer coefficient at elevated pressure and temperature and to ensure proper cooling and control NO_x emissions, the hp turbine inlet temperature was held to only 550°C compared to 825°C for the low pressure (lp) turbine (typical for a gas turbine without blade cooling). Moderate combustion inlet temperatures also facilitate the daily turbine starts needed for CAES operation [55]. Although the plant could operate at a lower heat rate if equipped with heat recuperators (to recover exhaust heat from the lp turbine for preheating the gas entering the hp turbine), this addition was omitted to minimize system start-up time [56,57].

McIntosh

Although high oil and gas prices through the early 1980s continued to draw the attention of utilities to CAES as a source for inexpensive peak power [48], not until a decade later did a CAES facility began operating in the United States. The 110 MW plant was built by the Alabama Electric Cooperative on the McIntosh salt dome in southwestern Alabama and has been in operation since 1991. It was designed for 26 hours of generation at full power and uses a single salt cavern (560,000 m³) designed to operate between 45 and 74 bar.

The project was developed by Dresser-Rand, but many of the operational aspects of the plant (inlet temperatures, pressures, etc.) are similar to those of the BBC design for the Huntorf plant. The McIntosh facility does, however, include a heat recuperator that reduces fuel consumption by approximately 22% at full load output and features a dual-fuel combustor capable of burning No. 2 fuel oil in addition to natural gas [1].

Although the plant experienced significant outages in its early operation, the causes were addressed through modifications of the high pressure combustor mounting and a redesign of the low pressure combustor [58]. These changes enabled the McIntosh plant, over 10 years of operation, to achieve 91.2 and 92.1% average starting reliabilities with 96.8 and 99.5% average running reliabilities for the generation and compression cycles, respectively [59].

Norton

A proposal has been under development to convert an idle limestone mine in Norton, Ohio into a storage reservoir for an 800 MW CAES facility (with provisional plans to expand to 2,700 MW [9 × 300 MW]. The mine, purchased in 1999, would provide 9.6 million cubic meters of storage and operate at pressures between 55 and 110 bar.

Iowa Stored Energy Park

The Iowa Association of Municipal Utilities (IAMU) is developing an aquifer CAES project in Dallas Center that will be directly coupled to a wind farm. The Iowa Stored Energy Park (ISEP), a 268 MW CAES plant coupled to 75 to 100 MW of wind capacity, was formally announced in December 2006. The CAES facility will occupy 40 acres within 30 miles of Des Moines and use a 3,000 foot deep anticline in a porous sandstone formation to store wind energy generated as far away as 100 to 200 miles from the site. This was the third location studied for ISEP after an initial screening of more than 20 geologic structures in the state. Studies of the chosen formation verified it has adequate size, depth, and caprock structure to support CAES operation [63].

Proposed Systems in Texas

Several factors make Texas and the surrounding region attractive for CAES development: First, the rapid growth of wind power in Texas (currently the largest and fastest growing wind market in the United States) imposes increasing burdens on existing load-following capacity in the region. Second, the considerable transmission bottlenecks and few inter-connection points with neighboring facilities present significant curtail-ment risks for wind developers as wind penetrations continue to increase. Finally, domal salt formations such as those used at the existing Huntorf and McIntosh sites exist in Texas. This geology has been proven to work well under CAES operating conditions and thus poses limited risk. Consequently, several parties have announced plans to develop CAES projects in Texas including a 540 MW (4 × 135 MW) system in Matagorda County based on the McIntosh Dresser-Rand design and utilizing a pre-viously developed brine cavern.

CAES Operation and Performance

Ramping, Switching and Part-Load Operation

The high part-load efficiency of CAES (see Figure 5.10) makes it well suited for balancing variable power sources such as wind. The heat rate increase at part-load is small relative to a conventional gas turbine because of the way the turbo expander output is controlled. Rather than changing the turbine inlet temperature as in a conventional turbine, the CAES output is controlled by adjusting the air flow rate with inlet temperatures kept constant at both expansion stages. This leads to better heat utilization and higher efficiency during part-load operation [56].

The McIntosh CAES plant delivers power at heat rates of 4330 kJ/kWh (LHV) at full load and 4750 kJ/kWh (LHV) at 20% load [58]. This excellent part-load behavior could be enhanced in modular systems such as the pro-posed Norton plant where the full output would be delivered by multiple modules. In this case, the system could ramp down to 2.2% of the full load output and still be within 10% of the full load output heat rate.

The ramp rates for a CAES system are also better than those of equiv-alent gas turbine plants. The McIntosh plant can ramp at approximately 18 MW per minute—about 60% greater than for typical gas turbines. The Matagorda plant proposed by Ridge Energy Storage is designed to bring its four 135 MW power train modules to full power in 14 minutes (or 7 minutes for an emergency start)—which translates to 9.6 to 19 MW per minute per module. These fast ramp rates together with efficient part load operation make CAES an ideal technology for balancing the stochastic variations in wind power.

To initiate compression operation, the turbine typically brings the machin-ery train to speed. After synchronization, the turbine is decoupled and shut off and the compressors are left operating. This means that the turbines are called upon to initiate both compression and generation. At the Huntorf

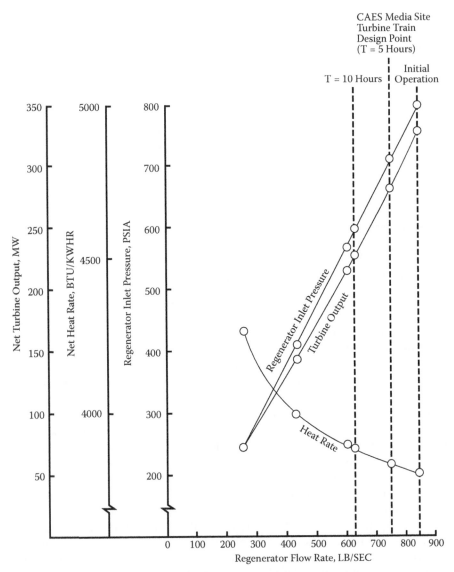

The above turbine output and heat rate values are as calculated and do not include margins.

FIGURE 5.10
Turbine performance characteristics for aquifer CAES based on EPRI design for Media, Illinois site (From Electric Power Research Institute, Compressed-Air Energy Storage Preliminary Design and Site Development Program in an Aquifer, EM-2351, November 1982.)

CAES facility, the switch from one operating mode to another is completely automated and requires a minimum of 20 minutes during which time the system is neither generating power nor compressing air [55]. The switchover time may have a significant impact for balancing rapid fluctuations in wind output. It is possible that alternative startup features such as use of an auxiliary startup motor could reduce this interval further [52].

Operation switchover time limitations may be eliminated altogether with new system designs that decouple the compression and turbo expander trains. By separating these components rather than linking them through a common shaft via a clutch as in the McIntosh and Huntorf systems, direct switching between compression and expansion operation is possible. This change also means compressor size can be optimized independently of the turbo expander design and permits standard production compressors to be used in the system configuration [57].

Constant Volume and Constant Pressure

A CAES can operate in a number of ways depending on the type of geology utilized for the storage reservoir. The most common mode is to operate the CAES under constant volume conditions. This means that the storage volume is a fixed, rigid reservoir operating over an appropriate pressure range.* This mode of operation offers two design options: (1) it is possible to design such a system to allow the hp turbine inlet pressure to vary with the cavern pressure (reducing output), or (2) keep the inlet pressure of the hp turbine constant by throttling the upstream air to a fixed pressure. Although this latter option requires a larger storage volume (due to throttling losses), it has been pursued at both the existing CAES facilities due to the increase in turbine efficiency attained for constant inlet pressure operation. The Huntorf CAES is designed to throttle the cavern air to 46 bar at the hp turbine inlet (with caverns operating at 48 to 66 bar) and the McIntosh system similarly throttles the incoming air to 45 bar (operating between 45 and 74 bar).

A third option is to keep the storage cavern at constant pressure throughout operation by using a head of water applied by an above-ground reservoir (see Figure 5.11). The use of compensated storage volumes minimizes losses and improves system efficiency, but care must be taken to manage flow instabilities in the water shaft such as the so-called champagne effect [64].

This technique is incompatible with salt-based caverns since a continual flow of water would dissolve the cavern walls. Brine cycling with a compensating column connected to a surface pond of saturated brine may be implemented, but biological concerns and groundwater contamination issues

---

* Although aquifer bubbles are not rigid bodies, the time scale at which the air–water interfaces migrate is much longer than CAES storage cycles and therefore porous rock systems can be approximated as fixed-volume air reservoirs in this context.

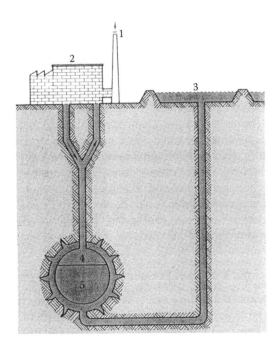

FIGURE 5.11
Constant pressure CAES reservoir with compensating water column. (1) Exhaust. (2) CAES plant. (3) Surface pond. (4) Stored air. (5) Water column. (From O. Weber, Air-storage gas turbine power station At Huntorf, *Brown Boveri Review*, 62: 332, 1975.)

would need to be addressed [56]. Since pressure compensated operation cannot be employed in aquifer systems, the use of constant-pressure CAES operation is limited to systems with reservoirs mined from hard rock.

Cavern Size

The energy storage density of CAES (depicted in Figure 5.12) depends on the maximum reservoir pressure, the storage volume operational mode, and the storage pressure ratio (see the appendix at the end of this chapter for derivation of relevant storage density equations). For all three cases considered in the appendix, the electric energy storage density E_{GEN}/V_S increases approximately linearly with increasing reservoir pressure p_{S2} (or equivalently with mass per unit volume $p_{S2} * M_W/RT_{S2}$). In some cases however, this may result in large heat loss in the aftercooler, depending on the thermal constraints of the cavern [65].

The use of a constant-pressure compensated cavern requires the smallest cavern by far. Zaugg estimates for a configuration similar to the Huntorf design (with a storage pressure of 60 bar) that a constant pressure cavern could deliver the same output with only 23% of the storage volume required for a constant volume configuration with variable inlet pressure ($p_{S2}/p_{S1} = 1.4$) [33]. If hard rock reservoirs are unavailable or too costly, pressure

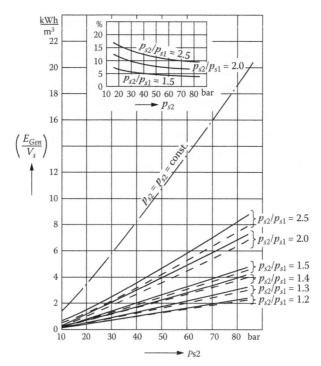

Fig. – Determining the size of the reservoir

E_{Gen}	= Generator energy
V_s	= Storage volume
P_{s2}	= Upper storage pressure
P_{s1}	= Lower storage pressure
—·—·	= Reservoir, case 1
——	= Reservoir, case 2
— —	= Reservoir, case 3
$T_{E_{HD}}$	= 825 °K, $T_{E_{ND}}$ = 1100 °K

FIGURE 5.12

Energy produced per unit volume for CAES with constant pressure reservoir (case 1), variable pressure reservoir (case 2), and variable pressure reservoir with constant turbine inlet pressure (case 3). Inset shows throttling losses associated with case 3 relative to variable inlet pressure scenario (case 2). (From P. Zaugg, Air-storage power generating plants, Brown *Boveri Review*, 62: 338, 1975.)

compensated systems will most likely not be options and a case 2 or case 3 design would be required.

Notably, although the throttling losses incurred in case 3 relative to the variable turbine inlet pressure system (case 2) implies a required larger storage volume, the penalty is not large (see Figure 5.12 inset). In particular, the throttling losses are small with large initial pressures ($p_{s2} > 60$ bar) and this is consistent with operations at all existing and proposed CAES facilities. Because this small penalty is offset by the benefits of higher turbine efficiency and simplified system operation, it is often optimal to operate a CAES in this mode (as is the case at both the Huntorf and McIntosh plants).

However, it may be advantageous to allow the inlet pressure to vary, depending on the geologic characteristics of a system. For aquifer systems, for example, due to the large amount of cushion gas needed, the storage pressure ratio p_{s2}/p_{s1} is relatively small (<1.5), such that the hp turbine can operate over the full storage reservoir pressure range with relatively small penalties relative to design point performance (see Figure 5.13) [52,55].

Although a variable pressure reservoir CAES system requires a larger storage volume than a compensated reservoir, volume requirements may be reduced substantially by appropriate design of the storage volume pressure range to an extent consistent with the pressure limits of the reservoir and the turbo machinery. The ratio of the energy storage density for hydraulic compensated storage to constant volume storage with inlet throttling (see Appendix) is given by

$$\left(1 - \left[\frac{p_{S1}}{p_{S2}}\right]^{\frac{1}{k_S}}\right) \tag{5.1}$$

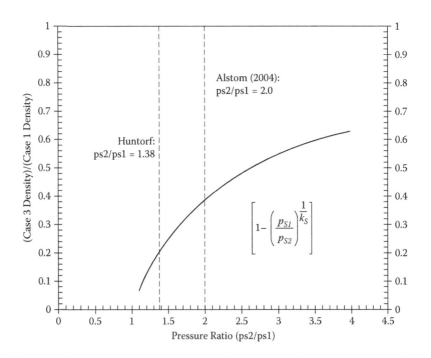

FIGURE 5.13
Ratio of storage energy density between constant volume CAES with constant turbine inlet pressure (case 3) and pressure compensated CAES reservoir (case 1) as a function of ratio of operating pressures of case 3 system (ps_2/ps_1). We assume ks = 1.4 and $(ps_2/Ts_2)/(ps_1/Ts_1) = 1$.

This term increases with p_{S2}/p_{S1} as shown in Figure 5.13. Thus, selecting formations that can accommodate large pressure swings and high maximum reservoir pressures will reduce land area requirements for CAES through increased storage energy density.

Typical numbers for E_{GEN}/V_S are 2 to 4 kWh/m^3 for lower pressure ratios such as those at Huntorf ($p_{S2}/p_{S1} = 1.38$; $p_{S2} = 66$ bar; $E_{GEN}/V_S = 3.74$) and 6 to 9 kWh/m^3 for the newer designs such one proposed by Alstom that features higher operating pressures and larger pressure ratios ($p_{S2}/p_{S1} = 2.0$; $p_{S2} = 110$ bar; $E_{GEN}/V_S = 8.44$) [1,66].

Performance Indices for CAES Systems

The energy performance of a conventional fossil fuel power plant is easily described by a single efficiency: the ratio of electrical energy generated to thermal energy in the fuel. The situation is more complicated for CAES due to the presence of two very different energy inputs. On the one hand, electricity is used to drive the compressors; natural gas or oil is burned to heat the air prior to expansion. This situation makes it difficult to describe CAES performance via a single index in a way that is universally useful—the most helpful single index depends on the specific application for CAES. Before turning to a discussion of alternative options for a single CAES performance index, we should consider two performance indices that apply to each energy input separately: the heat rate and the charging electricity ratio.

Heat Rate

The heat rate (HR) or fuel consumed per kilowatt hour of output for a CAES is a function of many system design parameters, but the design choice that most critically affects heat rate is the inclusion of a heat recovery system. The addition of a heat recuperator allows the system to capture the exhaust heat from the lp turbine to preheat the air withdrawn from the storage reservoir. Heat rates for CAES operations without heat recovery systems are typically 5500 to 6000 kJ/kWh LHV (e.g., 5870 kJ/kWh LHV for Huntorf). See Table 5.2. Heat rates with a recuperator are typically 4200 to 4500 kJ/kWh LHV (e.g., 4330 kJ/kWh for McIntosh). By comparison, a conventional gas turbine consumes at least twice this level of fuel (~9500 kJ/kWh LHV) because two thirds of the electrical output is used to run the compressor. Because the CAES compression energy is supplied separately, the system can achieve a much lower heat rate [1,56].

The addition of a heat recuperator reduced fuel consumption at McIntosh by 22% relative to operation without the component [58], but a high pressure combustor was still required. Newer CAES designs feature higher inlet temperatures at the lp turbine. The added heat generated at this stage facilitates the removal of the hp combustor from the design. In addition to further reducing

TABLE 5.2

Selected CAES Efficiency Expressions and Values Cited in Literature

Parameter	Definition	Reported Values	
		Without Heat Recuperator	With Heat Recuperator
Heat rate	$\eta_F = \dfrac{E_T}{E_F}$	6000 to 5500 kJ/kWh (~60 to 65%)	4500–4200 kJ/kWh (~80–85%)
Charging energy ratio	$\eta_{PE} = \dfrac{E_T}{E_M}$	1.2 to 1.4	1.4-1.6
Primary energy efficiency	$\eta_{PE} = \dfrac{E_T}{E_M/\eta_T + E_F}$	CAES charged from nuclear power ($\eta_T = 33\%$) [33]	
		24.5%	29.7%
		Charged from fossil fuel power plant ($\eta_T = 42\%$) [33]	
		28.2%	34.4%
		Charged from combined heat and power plant ($\eta_T = 35\%$) [65]	
		–	35.1 to 41.8%
		Charged from grid-averaged baseload power ($\eta_T = 35\%$; CER = 1.4) [70]	
			42–47%
Roundtrip Efficiency (1)	$\eta_{RT,1} = \dfrac{E_T}{E_M + \eta_{NG} E_F}$	4220 kJ LHV/kWh, CER = 1.5, $\eta_{NG} = 47.6\%$, [10]	
			81.7
Roundtrip Efficiency (2)	$\eta_{RT,2} = \dfrac{E_T - E_F\eta_{NG}}{E_M}$	4220 kJ LHV/kWh, Eo/Ei = 1.5, $\eta_{NG} = 38.2\%$, [74]	
			82.3%
Second Law Efficiency	$\eta_{II} = \dfrac{E_T}{E_{T,REV}}$	$T_O = 15$ C, $T_{MAX} = 900$ C, $p_S = 20$ bar [67]	
		58.7%	68.3%

fuel consumption, these systems also offer significant NO_x emission benefits relative to prior designs [66].

Charging Electricity Ratio

The second performance index for CAES is the ratio of generator output to compressor motor input—the charging electricity ratio (CER). Because of the fuel input, the CER exceeds unity and will typically lie in the range of 1.2 to 1.8 (kWh_{output}/kWh_{input}) [1,33,67]. The CER also takes into account piping and throttling losses and compressor and expander efficiencies. Throttling loss is a function of the reservoir pressure range. Turbine efficiency is especially important in the lp expansion stage, in which most of the enthalpy drop occurs and where approximately three quarters of the power is generated [68]. Increased turbine inlet temperatures (e.g., by using expander blade cooling technologies) would enhance the turbine and CAES electrical efficiencies as well [69].

Toward a Single CAES Performance Index

Several single-parameter performance indices have been proposed for CAES (see Table 5.2). The simplest is an efficiency η index defined as the ratio of energy generated by the turbine (E_T) to the sum of electrical energy delivered to the compressor motor (E_M) and the thermal energy in the fuel (E_F):

$$\eta = \frac{E_T}{E_M + E_F} \tag{5.2}$$

Typical HR and CER values of, respectively, 4220 kJ/kWh and 1.5 imply $\eta = 54\%$. However, because of the substantial difference between the energy qualities of the thermal energy in the fuel and the electrical energy supplied to the compressor, their sum is not a meaningful number. To estimate the total energy input to CAES, it is necessary to express both the fuel and compressor electricity on an equivalent energy basis. One approach is to express the electrical input as an equivalent quantity of thermal energy.

Primary Energy Efficiency

When CAES is used to convert baseload thermal power into peaking power (in place of gas turbines or other peaking units) one can introduce a primary energy efficiency η_{PE} defined in terms of the thermal efficiency of the baseload plant (η_T). Compressor motor energy input E_M is replaced by an expression for the effective thermal energy input required to produce E_M. Thus, the overall efficiency value reflects the system (grid + CAES) efficiency of converting primary (thermal) energy into electrical energy:

$$\eta_{PE} = \frac{E_T}{E_M/\eta_T + E_F} \tag{5.3}$$

This methodology has been applied to CAES units charged by nuclear and fossil fuel plants [33], CHP plants [65], and grid-averaged baseload power

[70]. Assuming $\eta_T = 40\%$ (as might characterize a modern supercritical steam electric plant) and the other parameters considered in the earlier calculation of η implies $\eta_{PE} = 35\%$.

In principle, this formulation of system efficiency can be applied to a wind/CAES by using the atmospheric efficiency of the wind turbines η_{WT} in place of the thermal plant efficiency η_T. This formulation proposed by Arsie et al. yields a system efficiency of 39% [71]. However, the use of atmospheric efficiency in this case does not serve the same function as using thermal efficiency. In the case of fossil fuel or nuclear power as the source of compressor energy, thermal efficiency provides a measure of the amount of primary fuel needed to deliver a quantity of electrical energy E_M. In contrast, the extraction of "fuel" in the case of wind energy does not affect the environment or overall cost of the plant. Consequently, this measure of the amount of atmospheric kinetic energy captured in providing E_M is not very helpful and in the case of CAES supporting renewables, is not the optimal formulation for CAES efficiency.

Round-Trip Efficiency

A CAES unit powered by wind energy may be compared to other electrical storage options that might be considered for wind back-up, for example, electrochemical or pumped hydroelectric storage. Such alternative storage systems are typically characterized by a round-trip electrical storage efficiency (η_{RT}) calculated as η_{RT} = (electricity output)/(electricity input). To facilitate comparisons of CAES to other electrical storage devices, a round-trip efficiency that employs an "effective" electricity input $\equiv E_M + \eta_{NG}{}^* E_F$ may be introduced. The second term is the amount of electricity that could be have been made from the natural gas input E_F, had that fuel been used to make electricity in a stand-alone power plant at efficiency η_{NG} instead of firing a CAES unit. The round-trip efficiency $\eta_{RT,1}$ so defined is

$$\eta_{RT,1} = \frac{E_T}{E_M + \eta_{NG} E_F} \tag{5.4}$$

This methodology has the advantage of providing an electricity-for-electricity round-trip storage efficiency that isolates the energy losses in the conversion of electricity to compressed air and back to electricity. Several values for η_{NG} have been proposed including the hypothetic Carnot cycle efficiency [67] and the efficiencies of commercial simple cycle and combined cycle power plants [10,72]. For typical natural gas power systems, (heat rates in the range 6700 to 9400 kJ/kWh), CAES round-trip efficiencies are in the range of 77 to 89%, assuming a 1.5 ratio of output to input electricity and a heat rate of 4220 kJ LHV/kWh. An exergy analysis of conventional CAES systems indicates that 47.6% of the fuel energy input is converted into electrical work [73]. For this measure of the thermal efficiency, the round-trip efficiency is 81.7%.

An alternative formulation $\eta_{RT,2}$ of an electrical round-trip storage efficiency introduces an output correction term $E_F * \eta_{NG}$. Instead of expressing the fuel input as an effective electrical input, the electrical output is adjusted by subtracting the assumed contribution to the output attributable to the fuel. Correspondingly the output attributable to the electrical input is $E_T - E_F * \eta_{NG}$ [74].

$$\eta_{RT,2} = \frac{E_T - E_F \eta_{NG}}{E_M} \quad (5.5)$$

However, this interpretation is a bit different since the term in the numerator attempts to estimate the residual output of the plant not attributed to natural gas output. Therefore, the conversion efficiency factor serves to separate this component of the output that would otherwise be generated in a conventional turbine from the overall output of the CAES unit. The efficiency of a stand-alone natural gas fired combustion turbine (38%) is a more appropriate benchmark than an exergy-based measure of the CAES system fuel conversion. Using the same assumptions as for $\eta_{RT,1}$ and a CT conversion efficiency, the round-trip efficiency is 82.8% which is consistent with the measure derived above. Thus, depending on the index chosen for its measure, the round-trip efficiency for CAES is ~82%.—in the same range as the round-trip efficiencies cited for other bulk energy storage technologies such as pumped hydroelectric storage (74%) and vanadium flow batteries (75%) [72].

Additional Approaches

Still another measure of the efficiency of CAES proposed by Schainker et al. may be useful for an economic evaluation of CAES in load leveling or arbitrage applications. This approach is similar to $\eta_{RT,1}$ in that it adjusts the fuel input by a correction factor:

$$\eta_{AD} = \frac{E_T}{E_F/CR + E_M} \quad (5.6)$$

In this case, however, the fuel input is converted to equivalent electricity not by using the primary energy conversion efficiency for natural gas, but rather by using the cost ratio CR ≡ (off-peak electricity price)/(fuel price) [75]. Although this index may be helpful for deciding how to operate a given CAES unit over time, it varies significantly over both time and geographical region and so is not a useful general plant characterization. A final description of CAES efficiency compares CAES output to the output of a thermodynamically ideal CAES plant operating between ambient temperature T_o and T_{max} [67]:

$$\eta_{II} = \frac{E_T}{E_{T,REV}} \tag{5.7}$$

$$E_{T,REV} = E_M + E_F - T_o \cdot \Delta S = E_M + E_F - T_o \cdot E_F/T_{MAX} \tag{5.8}$$

Analysis of a conventional CAES system yields a second law efficiency of $\eta_{II} = 68\%$ with a recuperator and 59 to 61% without one.*

Ultimately, the choice of efficiency measure remains an open question because thermal energy and electrical energy quantities cannot be combined by algebraic manipulation. The formulations provided in this section provide only a basis for comparison with other storage technologies. As indicated above, the relevant expression is determined in large part by each specific application.

Advanced Technology Options

Although commercial CAES plants have been operating for several decades, the technology is still in an early stage of development. This is reflected in the fact that the two existing plants are based largely on conventional gas turbine and steam turbine technologies. Consequently, various technological improvements may eventually enhance performance and reduce costs over relatively few product cycles.

One option that has attracted interest is to reduce (and perhaps eliminate) the CAES fuel requirements and associated GHG emissions by recovery and storage of the high-quality heat of compression in thermal energy storage (TES) systems. Heat recovery may be implemented at some or all compression stages, thus allowing stored heat to be used in place of fuel to reheat air withdrawn from the CAES cavern, thereby partially or completely

---

* The range of efficiencies for a system without a recuperator reflects changes in system performance due to varying storage pressures (p_S = 20 to 70 bar). The change in efficiency was <1% for a system with a recuperator.

eliminating the need for natural gas [67]. To be economically feasible, fuel cost reductions must offset the additional capital cost associated with a TES system. Early studies found that very high fuel prices would be required to justify such systems, making adiabatic CAES too costly for commercial use [76–80].

More recent studies, however, suggest that new TES technologies along with improvements in compressor and turbine systems may make the so-called advanced adiabatic CAES (AA-CAES) technology economically viable (see Table 5.3) [17,81]. One such AA-CAES concept with a high efficiency turbine and a high capacity TES, achieves a round-trip efficiency of approximately 70% with no fuel consumption (see Figure 5.14) [39]. But it should be noted that the efficiency gain of adiabatic systems over multistage compression with intercooling is small [82], and both the fuel use and GHG emissions for wind/CAES systems are already very modest [10].

Another proposal is to use biomass-derived fuels to reheat the air withdrawn from storage. This could reduce GHG emissions and decouple plant economics from fuel price fluctuations [83]. It may also allow CAES to be run on fuel produced locally, thereby facilitating the use of energy crops in remote, wind-rich areas and eliminating the need for natural gas supplies. However, as in the adiabatic case, the emissions benefit would be small because the emissions level of wind/CAES is already quite low (~2/3 the rate for a coal IGCC plant with CCS) [10]. Moreover, a biofuel plant dedicated to a wind/CAES system would require fuel storage because biofuels must be produced in large-scale plants that run flat-out in order to be cost effective, while CAES expander capacity factors for backing wind are typically modest [10].

A CAES variant proposed for wind applications is to replace the electrical generator in a wind turbine nacelle with a compact compressor. This would enable the wind turbine to generate compressed air directly, thereby eliminating two energy conversion processes. However, the reduced losses and potential drop in turbine capital cost would have to offset the added capital cost of the compact compressors and the considerable cost of the high pressure piping network needed to transport the compressed air from each turbine to the storage reservoir.

In contrast to the option of coupling intermittent wind to CAES to obtain baseload electricity, CAES might also be coupled to baseload power systems to facilitate the use of such systems to provide load-following and/or peaking power—the function originally envisioned for CAES—e.g., by coupling CAES to a coal IGCC plant [84,85].

Improving CAES turbo machinery is a promising area for innovation. CAES turbine operating temperatures might be increased, thereby increasing their efficiency by introducing turbine blade cooling technologies routinely deployed in conventional gas turbines but not in commercial CAES units. Other advanced CAES concepts include various humidification and steam injection schemes that may be used to boost the power output of a

TABLE 5.3
Main Thermal Energy Storage (TES) Concepts for AA-CAES [81][a]

	Solid TES					Liquid TES		
Concept	Rock Bed	Cowper Derivative	Concrete Walls	Cast Iron Slabs	Hybrid Phase-Change Materials	Two-Tank	One-Tank Thermocline	Air–Liquid
Contact	Direct	Direct	Direct	Direct	Direct	Indirect	Indirect	Indirect
Storage material	Natural stone	Ceramic	Concrete	Cast iron	Ceramic, salt	Nitrate salt, mineral oil	Nitrate salt, mineral oil	Nitrate salt, mineral oil

[a] Storage technologies chosen on basis of ability to deliver 120 to 1200 MWh (thermal), maintain high consistency of outlet temperature, and cover full temperature range (50 to 650°C).

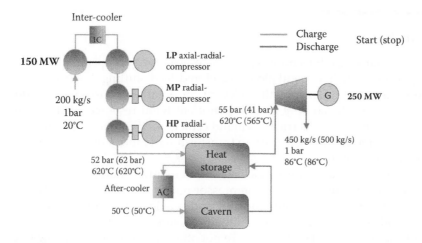

FIGURE 5.14
Possible technical concept for AA-CAES system under development. (From B. Calaminus, Innovative adiabatic compressed air energy storage system of EnBW in Lower Saxony, Second International Renewable Energy Storage Conference, Bonn, 2007.)

system and reduce storage requirements [86]. The CAES combined cycle is still another option that allows a system to generate electricity even when the compressed air storage reservoir is depleted [87,88].

A recent hybrid CAES system design incorporates a standard combustion turbine in place of the turbo expander chain in a traditional CAES design. The air withdrawn from storage is heated by a recuperator at the turbine exhaust instead of heating by fuel combustors as in a conventional CAES plant. The heated air is then injected into the turbine to boost the output. The use of commercial technology and the elimination of fuel combustors could reduce the capital cost substantially and provide a low risk option for early adoption of bulk storage. Such an air injection CAES (AI-CAES) plant could also include a bottoming cycle and TES system to reduce fuel consumption further [57,89].

Although it is possible that new CAES concepts will bring important changes to the way air storage operates and the way wind power is stored, performance and cost gains are most likely to arise in the near term as a result of marginal improvements in existing CAES designs as a result of learning by doing. Thus, after technology launch in the market, costs for new technologies such as CAES can be expected to decline at faster rates than for mature technologies and more quickly based on faster rates of deployment. This phenomenon bodes well for wind/CAES as a baseload power climate change mitigation option if there is a way to obtain substantial early market experience.

Conclusions

Traditionally, CAES technology has been used for grid operational support applications such as regulation control and load shifting, but a major new opportunity that is especially relevant for a carbon constrained world is to achieve exploitation of large intermittent wind resources that are often remote from major electricity demand centers. CAES appears to have many of the characteristics necessary to transform wind into a mainstay of global electricity generation.

Backing wind to produce baseload output requires short response times to accommodate fluctuations in compressor power and turbine load. The ability of a CAES to ramp output quickly and provide efficient part-load operation makes it particularly well suited for balancing such fluctuations—key performance characteristics that are not often called upon at existing CAES plants that simply store low-cost off-peak electricity for use when electricity is more valuable.

Air storage volume requirements translate into a geologic footprint comprising ~15% of the wind farm land area, so that CAES will exert relatively limited impact on land use and ecology. The wide availability of potentially suitable geology in wind-rich areas points to CAES as a suitable technology for making baseload power from wind—thereby making it feasible to provide wind power at electric grid penetrations far greater than 20% rates that are feasible without storage. And, to the extent that wind-rich regions are remote from major electricity markets, such baseload power can often be delivered to distant markets via high voltage transmission lines at attractive costs.

Aquifer CAES seems to be the most suitable storage geology for wind/CAES in the United States based on the potential for low development costs and because regions with porous rock geologies are strongly correlated with the onshore wind-rich regions of the country. Aquifer CAES technology has been studied for nearly three decades, but the first commercial plant was formally announced only recently. Nevertheless, a great deal of commercial experience can be gleaned from the natural gas storage industry that uses geologies similar to those needed for CAES to meet seasonal heating demand fluctuations. The methodologies for evaluating natural gas storage reservoirs have proven directly applicable to aquifer CAES development, but several differences in using methane and air as storage fluids must be taken into account. Care must be taken to carefully characterize local mineralogy, existing bacterial populations, and relevant corrosion mechanisms to anticipate and prevent problems resulting from the introduction of air into porous underground media. Methods for mitigating the impacts of these factors such as air dehydration, particulate filtration, and biocide application could help expand the number of suitable sites. Despite the various issues that must be considered, they do not obviously diminish CAES as a strong candidate for large scale energy storage.

Quantifying the full deployment potential of CAES will require more detailed characterizations of existing porous rock formations and operational experience from multiple plants operating over a wide variety of geologic conditions.

References

1. Electric Power Research Institute and U.S. Department of Energy. 2003. *Handbook of Energy Storage for Transmission and Distribution Applications.* Palo Alto, CA, and Washington.
2. Electric Power Research Institute and U.S. Department of Energy. 2004. *Energy Storage for Grid-Connected Wind Generation Applications.* Palo Alto, CA, and Washington.
3. S. M. Schoenung. 2001. Characteristics and Technologies for Long-vs. Short-Term Energy Storage: A Study by the DOE Energy Storage Systems Program. Sandia National Laboratories, Albuquerque, New Mexico.
4. F. R. McLarnon and E. J. Cairns. 1989. Energy storage. In *Annual Review of Energy,* Vol. 14, p. 241.
5. A. Gonzalez, B. Ó. Gallachóir, E. McKeogh et al. May 2004. Study of Electricity Storage Technologies and Their Potential to Address Wind Energy Intermittency in Ireland. Department of Civil and Environmental Engineering, University College Cork.
6. L. W. M. Beurskens, M. D. Noord, and A. F. Wals. December 2003. Analysis in the Framework of the Investire Network: Economic Performance of Storage Technologies, Energy Research Centre of the Netherlands, Petten.
7. Cambridge Energy Research Associates. 2002. Energy Storage: An Emerging Competitor in the Distributed Energy Industry. Cambridge, MA.
8. Electric Power Research Institute. 2005. Wind Power Integration: Energy Storage for Firming and Shaping, Palo Alto, CA.
9. A. J. Cavallo, 1995. High-capacity factor wind energy systems, *Journal of Solar Energy,* 117: 137.
10. J. B. Greenblatt, S. Succar, D. C. Denkenberger et al. 2007. Baseload wind energy: modeling the competition between gas turbines and compressed air energy storage for supplemental generation, *Energy Policy,* 35: 1474.
11. J. F. DeCarolis and D. W. Keith, 2006. The economics of large-scale wind power in a carbon constrained world. *Energy Policy,* 34: 395.
12. I. Arsie, V. Marano, G. Rizzo et al. 2006. Energy and Economic Evaluation of a Hybrid Power Plant with Wind Turbines and Compressed Air Energy Storage. *ASME Power Conference*, Atlanta, GA.
13. P. Denholm, G. L. Kulcinski, and T. Holloway. 2005. Emissions and energy efficiency assessment of baseload wind energy systems, *Environmental Science and Technology,* 39: 1903.
14. A. Cavallo. 2007. Controllable and affordable utility-scale electricity from intermittent wind resources and compressed air energy storage (CAES). *Energy,* 32: 120.

15. N. Desai, S. Nelson, S. Garza et al. August 21, 2003. Study of Electric Transmission in Conjunction with Energy Storage Technology, Lower Colorado River Authority, Texas State Energy Conservation Office, Austin.
16. N. Desai, S. Gonzalez, D. J. Pemberton et al. June 27, 2005. Economic Impact of CAES on Wind in Texas, Oklahoma, and New Mexico. Texas State Energy Conservation Office, Austin.
17. G. Salgi and H. Lund. 2006. Compressed air energy storage in Denmark: a feasibility study and an overall energy system analysis. World Renewable Energy Congress, Florence.
18. D. J. Swider. 2007. Compressed air energy storage in an electricity system with significant wind power generation. *IEEE Transactions on Energy Conversion*, 22: 95.
19. D. L. Katz and E. R. Lady. 1976. *Compressed Air Storage for Electric Power Generation*. Ulrich, Ann Arbor, MI.
20. Z. S. Stys. 1977. Compressed air storage for load leveling of nuclear power plants. In *Proceedings of 12th Intersociety Energy Conversion Engineering Conference*, vol. 2. American Nuclear Society, Washington, p. 1023.
21. K. G. Vosburgh, D. C. Golibersuch, P. M. Jarvis et al. 1977. Compressed air energy storage for electric utility load leveling. In *Proceedings of 12th Intersociety Energy Conversion Engineering Conference* vol. 2. American Nuclear Society, Washington, p. 1016.
22. B. Sorensen. 1976. Dependability of wind energy generators with short-term energy storage. *Science*, 194: 935.
23. H. Holttinen, B. Lemström, P. Meibom et al. 2007. Design and Operation of Power Systems with Large Amounts of Wind Power: State-of-the Art Report. VTT Technical Research Centre, Vuorimiehentie, Finland.
24. EnerNex Corporation. May 22, 2006. Wind Integration Study for the Public Service Company of Colorado.
25. E. A. DeMeo, G. A. Jordan, C. Kalich et al. 2007. Accomodating wind's natural behavior. *IEEE Power and Energy*, 5: 59.
26. R. DeCorso, L. Davis, D. Horazak et al. 2006. Parametric study of payoff in applications of air energy storage plants: an economic model for future applications. PowerGen International Conference, Orlando, FL.
27. International Energy Agency. 2006. World Energy Outlook 2006. Paris.
28. J. B. Greenblatt. October 1, 2005. Wind as a Source of Energy Now and in the Future. Interacademy Council, Amsterdam.
29. D. L. Elliott, L. L. Wendell, and G. L. Gower. August 1991. Assessment of Available Windy Land Area and Wind Energy Potential in the Contiguous United States. U.S. Department of Energy, Pacific Northwest Laboratory, Richland, WA. PNL-7789, DE91 018887.
30. A. J. Cavallo and M. B. Keck. 1995. Cost Effective Seasonal Storage of Wind Energy. Houston, p. 119.
31. R. Wiser and M. Bolinger. May 2007. Annual Report on U.S. Wind Power Installation, Costs, and Performance Trends, 2006. U.S. Department of Energy, Washington, 02007-2433.
32. A. Ter-Gazarian. 1994. *Energy Storage for Power Systems*. Redwood Books, Trowbridge, UK.
33. P. Zaugg, 1975. Air storage power generating plants. *Brown Boveri Review*, 62: 338.
34. K. Allen. 1985. CAES: the underground portion, *IEEE Transactions on Power Apparatus and Systems*, 104: 809.

35. B. Mehta. 1992. CAES geology. *EPRI Journal,* 17: 38.
36. F. Crotogino, K. U. Mohmeyer, and R. Scharf. 2002. Huntorf CAES: More Than Twenty Years of Successful Operation. Solution Mining Research Institute Meeting Orlando, FL.
37. W. F. Adolfson, J. S. Mahan, E. M. Schmid et al. 1979. Geologic Issues Related to Underground Pumped Hydroelectric and Compressed Air Energy Storage. 14th Intersociety Energy Conversion Engineering Conference, Boston.
38. K. L. DeVries, K. D. Mellegard, G. D. Callahan et al. 2005. Roof Stability for Natural Gas Storage in Bedded Salt. U.S. Department of Energy, National Energy Technology Laboratory Topical Report RSI-1829, DE-FG26-02NT41651.
39. B. Calaminus. 2007. Innovative Adiabatic Compressed Air Energy Storage System of EnBW in Lower Saxony. Second International Renewable Energy Storage Conference, Bonn.
40. K. Sipila, M. Wistbacka, and A. Vaatainen. 1994. Compressed air energy storage in an old mine. *Modern Power Systems,* 14: 19.
41. S. Shepard and S. van der Linden. 2001. Compressed air energy storage adapts proven technology to address market opportunities. *Power Engineering,* 105: 34.
42. M. Schwartz and D. Elliott. 2001. Remapping of the Wind Energy Resource in the Midwestern United States. Third Symposium on Environmental Applications, Annual Meeting of American Meteorological Society, Orlando, FL.
43. M. Schwartz and D. Elliott. 2004. Validation of updated state wind resource maps for the United States. In *Proceedings of the World Renewable Energy Congress VIII,* Denver. Elsevier, Amsterdam.
44. D. Elliott and M. Schwartz. 2002. Validation of new wind resource maps. In *Conference Proceedings of American Wind Energy Association, WindPower 2002 Conference,* Portland, OR.
45. B. R. Mehta. 1990. Siting compressed air energy storage plants. In *Proceedings of American Power Conference,* Chicago, p. 73.
46. Y. Zimmels, F. Kirzhner, and B. Krasovitski. 2003. Energy loss of compressed air storage in hard rock. In *Fourth International Conference on Ecosystems and Sustainable Development,* Siena, Italy, p. 847.
47. T. Brandshaug and A. F. Fossum. 1980. Numerical studies of CAES caverns in hard rock. In *Mechanical, Magnetic, and Underground Energy Storage Annual Contractors' Review,* Washington, p. 206.
48. N. Lihach. 1982. Breaking new ground with CAES. *EPRI Journal,* 7: 17.
49. Electric Power Research Institute. 1990. Compressed Air Energy Storage Using Hard Rock Geology: Test Facility and Results. Palo Alto, CA.
50. Federal Energy Research Commission. September 30, 2004. Current State of and Issues Concerning Underground Natural Gas Storage. AD04-11-000.
51. Electric Power Research Institute. August 1994. Evaluation of Benefits and Identification of Sites for a CAES Plant in New York State. Palo Alto, CA, TR-104268.
52. Electric Power Research Institute. November 1982. Compressed-Air Energy Storage: Preliminary Design and Site Development Program in an Aquifer. Palo Alto, CA, EM-2351.
53. Electric Power Research Institute. 1990. Compressed Air Energy Storage: Pittsfield Aquifer Field Test. Palo Alto, CA, GS-6688.
54. S. van der Linden. 2006. Bulk energy storage potential in the USA: current developments and future prospects. *Energy,* 31: 3446.

55. A. J. Karalis, E. J. Sosnowicz, and Z. S. Stys. 1985. Air storage requirements for 220 MWe CAES plant as function of turbomachinery selection and operation. *IEEE Transactions on Power Apparatus and Systems*, 104: 803.

56. O. Weber. 1975. Air storage gas turbine power station at Huntorf. *Brown Boveri Review*, 62: 332.

57. S. van der Linden. 2007. Review of CAES systems development and current innovations. In *Electrical Energy Storage Applications and Technology Conference*, San Francisco.

58. V. De Biasi. 1998. The 110 MW McIntosh CAES plant: over 90% availability and 95% reliability. *Gas Turbine World*, 28: 26.

59. L. Davis and R. Schainker. 2006. Compressed air energy storage (CAES): Alabama Electric Cooperative McIntosh Plant overview and operational history. In *Electricity Storage Association Meeting: Energy Storage in Action*, Knoxville, TN.

60. Ohio Power Siting Board. March 20, 2001. Staff Report of Investigation and Recommended Findings. Public Utilities Commission of Ohio, Columbus.

61. Ohio Power Siting Board. March 8, 2006. Staff Investigation Report and Recommendation. Public Utilities Commission of Ohio, Columbus.

62. J. A. Strom. February 22, 2007. Norton Energy Storage: Annual Project Progress Status Report to Ohio Power Siting Board Staff.

63. Iowa Association of Municipal Utilities. 2006. Site for ISEP development is officially announced. *IAMU Newsletter*, p. 1.

64. A. J. Giramonti and E. B. Smith. 1981. Control of Champagne Effect in CAES Power Plants. Atlanta, p. 984.

65. B. Elmegaard, N. Szameitat, and W. Brix. 2005. Compressed air energy storage (CAES) possibilities in Denmark. 18th International Conference on Efficiency, Cost, Optimization, Simulation and Environmental Impact of Energy Systems, Trondheim.

66. I. Tuschy, R. Althaus, R. Gerdes et al. 2004. Evolution of gas turbines for compressed air energy storage. *VGB Powertech*, 85: 84.

67. E. Macchi and G. Lozza. 1987. Study of thermodynamic performance of CAES plants including unsteady effects. In *Gas Turbine Conference and Exhibition*, Anaheim, CA, p. 10.

68. D. R. Hounslow, W. Grindley, R. M. Loughlin et al. 1998. Development of a combustion system for a 110 MW CAES plant. *Journal of Engineering for Gas Turbines and Power: Transactions of ASME*, 120: 875.

69. I. Tuschy, R. Althaus, R. Gerdes et al. 2002. Compressed air energy storage with high efficiency and power output, *VDI Berichte*, p. 57.

70. Y. S. H. Najjar and M. S. Zaamout. 1998. Performance analysis of compressed air energy storage (CAES) plant for dry regions. *Energy Conversion and Management*, 39: 1503.

71. I. Arsie, V. Marano, G. Nappi, et al. 2005. A model of a hybrid power plant with wind turbines and compressed air energy storage. In *ASME Power Conference*, Chicago, p. 987.

72. P. Denholm and G. L. Kulcinski. June 2003. Net Energy Balance and Greenhouse Gas Emissions from Renewable Energy Storage Systems. Energy Center of Wisconsin, Madison.

73. P. Zaugg. 1985. Energy flow diagrams for diabatic air storage plants. *Brown Boveri Review*, 72: 179.

74. P. Denholm and G. L. Kulcinski. 2004. Life cycle energy requirements and greenhouse gas emissions from large scale energy storage systems. *Energy Conversion and Management,* 45: 2153.

75. R. Schainker, M. Nakhamkin, J. R. Stange et al. 1984. *Turbomachinery Engineering and Optimization for 25 and 50 MW Compressed Air Energy Storage Systems.* Elsevier, Amsterdam.

76. I. Glendenning. 1981. Compressed air storage, *Physics in Technology,* 12: 103.

77. D. Kreid. 1978. Analysis of advanced compressed air energy storage concepts. In *ASME Thermophysics and Heat Transfer Conference,* Palo Alto, CA, p. 11.

78. S. C. Schulte, 1979. Economics of thermal energy storage for compressed air energy storage systems. In *Proceedings of Mechanical and Magnetic Energy Storage Review Meeting,* Washington, p. 191.

79. R. B. Schainker, B. Mehta, and R. Pollak. 1993. *Overview of CAES Technology.* Chicago, p. 992.

80. R. W. Reilly and D. R. Brown. 1981. Comparative economic analysis of several CAES design studies. In *ASME Proceedings of Intersociety Energy Conversion Engineering Conference,* Atlanta, p. 989.

81. C. Bullough, C. Gatzen, C. Jakiel et al. 2004. Advanced adiabatic compressed air energy storage for integration of wing energy. In *European Wind Energy Conference,* London.

82. S. Succar and R. H. Williams. February 2008. Compressed Air Energy Storage: Theory, Operation and Applications. Princeton Environmental Institute, Princeton, NJ.

83. P. Denholm. 2006. Improving the technical, environmental and social performance of wind energy systems using biomass-based energy storage. *Renewable Energy,* 31: 1355.

84. M. Nakhamkin, M. Patel, E. Swensen et al. 1991. Application of air saturation to integrated coal gasification. CAES Power Plant Concepts, San Diego, CA.

85. M. Nakhamkin, M. Patel, L. Andersson et al. 1991. Analysis of integrated coal gasification system. CAES Power Plant Concepts, San Diego, CA.

86. M. Nakhamkin, E. Swensen, P. Abitante et al. 1992. Technical and economic characteristics of compressed air energy storage concepts with air humidification. In *Second International Conference on Compressed Air Energy Storage,* San Francisco.

87. K. Yoshimoto and T. Nanahara. 2005. Optimal daily operation of electric power systems with an ACC-CAES generating system. *Denki Gakkai Ronbunshi (Japan),* 152: 15.

88. M. Nakhamkin, E. Swensen, R. Schainker et al. 1991. Compressed air energy storage: survey of advanced CAES development. CAES Power Plant Concepts, San Diego, CA.

89. M. Nakhamkin. 2006. Novel compressed air energy storage concepts. In *Energy Storage Association Meeting: Energy Storage in Action,* Knoxville, TN.

Appendix: Storage Volume Requirement

One of the keys to assessing the geologic requirements for CAES is to understand how much electrical energy can be generated per unit volume of storage cavern capacity (E_{GEN}/V_S). The electrical output of the turbine (E_{GEN}) is given by

$$E_{GEN} = \eta_M \eta_G \int_0^t \dot{m}_T\, w_{CV,TOT}\, dt \tag{5.9}$$

where the integral is the mechanical work generated by the expansion of air and fuel in the turbine, $w_{CV,TOT}$ = total mechanical work per unit mass generated in this process, $\dot{m}_T$ = air mass flow rate, t = time required to deplete a full storage reservoir at full output power, η_M = mechanical efficiency of the turbine (which reflects turbine bearing losses), and η_G = electric generator efficiency.

Since all CAES systems to date are based on two expansion stages, the work output can be expressed as the sum of the output from the two stages. The first term reflects the work output from the hp turbine that expands the air from the hp turbine inlet pressure (p_1) to the lp turbine inlet pressure (p_2). Likewise, the second term reflects the expansion work derived from the expansion from p_2 to barometric pressure (p_b).

$$w_{CV,TOT} = w_{CV1} + w_{CV2} = -\int_{p_1}^{p_2} v\,dp - \int_{p_2}^{p_b} v\,dp \tag{5.10}$$

Consider first the work output from the first expansion stage. Assuming adiabatic compression and that the working fluid is an ideal gas with a constant specific heat (so that $P \cdot v^k = c$, a constant, where $k_1 \equiv C_{p1}/C_{v1}$), the work per unit mass is

$$w_{CV1} = \int_{p_2}^{p_1} v\,dp = c^{1/k_1} \int_{p_2}^{p_1} \frac{dp}{p^{1/k_1}} \tag{5.11}$$

$$= \frac{k_1}{k_1 - 1}\left[p\left(\frac{c}{p}\right)^{1/k_1} \right]_{p_2}^{p_1} = \frac{k_1}{k_1 - 1}\left[p_1 v_1 - p_2 v_2 \right] \tag{5.12}$$

$$= \frac{c_v}{c_p - c_v}\frac{c_p}{c_v} p_1 v_1\left(1 - \frac{p_2 v_2}{p_1 v_1} \right) \tag{5.13}$$

$$= c_p T_1 \left[1 - \left(\frac{p_2}{p_1} \right)^{\frac{k_1 - 1}{k_1}} \right] \tag{5.14}$$

Combining with a similar expression for the second stage gives the total work per unit mass for the process ($w_{CV,TOT}$):

$$w_{CV,TOT} = c_{p2} T_2 \left(\frac{c_{p1} T_1}{c_{p2} T_2} \left[1 - \left(\frac{p_2}{p_1} \right)^{\frac{k_1 - 1}{k_1}} \right] + \left[1 - \left(\frac{p_b}{p_2} \right)^{\frac{k_2 - 1}{k_2}} \right] \right) \tag{5.15}$$

Furthermore, the total mass flow through the turbine can be expressed as separate air and fuel input terms:

$$\dot{m}_T = \dot{m}_A + \dot{m}_F = \dot{m}_A \left(1 + \frac{\dot{m}_F}{\dot{m}_A} \right) \tag{5.16}$$

Since

$$\frac{\dot{m}_F}{\dot{m}_A} \approx \text{constant} \tag{5.17}$$

the result is

$$\frac{E_{GEN}}{V_S} = \frac{\alpha}{V_S} \int_0^t \dot{m}_A \left(\beta + 1 - \left(\frac{p_b}{p_2} \right)^{\frac{k_2 - 1}{k_2}} \right) dt \tag{5.18}$$

where

$$\alpha = \eta_M \, \eta_G \, c_{p2} T_2 \left(1 + \frac{\dot{m}_F}{\dot{m}_A} \right) \tag{5.19}$$

and

$$\beta = \frac{c_{p1} T_1}{c_{p2} T_2} \left[1 - \left(\frac{p_2}{p_1} \right)^{\frac{k_1 - 1}{k_1}} \right] \tag{5.20}$$

Case 1: Constant Cavern Pressure

First, consider the case of a CAES system with constant cavern pressure such as a hard rock cavern with hydraulic compensation (see Figure 5.12). The mass flow of air is constant throughout the process and can be expressed as a simple ratio:

$$\dot{m}_A = \frac{m_A}{t} = \frac{p_S V_S M_W}{R T_S t} \tag{5.21}$$

Likewise, since the inlet pressures and temperatures are constant in time, Equation (5.30) reduces to

$$\frac{E_{GEN}}{V_S} = \frac{\alpha}{V_S} \dot{m}_A \left(\beta + 1 - \left(\frac{p_b}{p_2} \right)^{\frac{k_2-1}{k_2}} \right) \int_0^t dt \tag{5.22}$$

Combining these expressions,

$$\frac{E_{GEN}}{V_S} = \frac{\alpha}{RT_S} \frac{M_W}{p_S} \left(\beta + 1 - \left(\frac{p_b}{p_2} \right)^{\frac{k_2-1}{k_2}} \right) \tag{5.23}$$

Case 2: Variable Cavern Pressure and Variable Turbine Inlet Pressure

In the case of a variable pressure CAES system, the pressure at the turbine inlet is allowed to vary over the operating range of the storage volume (from p_{S2} to p_{S1}). However, since the pressure ratio across the hp turbine (p_2/p_1) remains constant, the pressure ratio across the lp turbine is proportional to the cavern pressure p_S [33]:

$$\frac{p_b}{p_2} = \frac{p_1}{p_2} \frac{p_b}{\varphi p_S} = \frac{\text{constant}}{p_S} \tag{5.24}$$

where φ is a correction factor that accounts for the pressure loss from the storage reservoir to the turbine inlet (~0.90).

$$\dot{m}_A = \frac{d}{dt}\left(\frac{V_S p_S\, M_W}{R T_S}\right) = \frac{d}{dt}\left[\frac{V_S M_W p_S}{R T_{S2}}\left(\frac{p_{S2}}{p_S}\right)^{\frac{k_S-1}{k_S}}\right] \tag{5.25}$$

$$\dot{m}_A = \frac{1}{k_S}\left[\frac{V_S\, M_W}{R T_{S2}}\left(\frac{p_{S2}}{p_S}\right)^{\frac{k_S-1}{k_S}}\right]\frac{dp_S}{dt} \tag{5.26}$$

Substituting Equations (5.36) and (5.38) into (5.30), the energy storage density is

$$\frac{E_{GEN}}{V_S} = \frac{\alpha\, M_W}{R T_{S2}}\frac{p_{S2}^{\frac{k_S-1}{k_S}}}{k_S}\int_{p_{S1}}^{p_{S2}}\left(\frac{1}{p_S}\right)^{\frac{k_S-1}{k_S}}\left(\beta+1-\left(\frac{p_1}{p_2}\frac{p_b}{\varphi p_S}\right)^{\frac{k_2-1}{k_2}}\right)dp_S \tag{5.27}$$

$$= \frac{\alpha\, M_W}{R T_{S2}}\frac{p_{S2}^{\frac{k_S-1}{k_S}}}{k_S}\left\{(\beta+1)\int_{p_{S1}}^{p_{S2}}\left(\frac{1}{p_S}\right)^{\frac{k_S-1}{k_S}}dp_S -\left(\frac{p_1}{p_2}\frac{p_b}{\varphi}\right)^{\frac{k_2-1}{k_2}}\int_{p_{S1}}^{p_{S2}}\left(\frac{1}{p_S}\right)^{\frac{k_S-1}{k_S}+\frac{k_2-1}{k_2}}dp_S\right\} \tag{5.28}$$

$$= \frac{\alpha\, M_W p_{S2}}{R T_{S2}\, k_S}\left\{(\beta+1)\frac{1}{p_{S2}^{1/k_S}}\int_{p_{S1}}^{p_{S2}}p_S^{\frac{1}{k_S}-1}dp_S -\left(\frac{p_1}{p_2}\frac{p_b}{\varphi}\right)^{\frac{k_2-1}{k_2}}\frac{p_{S2}^{\frac{k_2-1}{k_2}}}{p_{S2}^{\frac{1}{k_S}+\frac{1}{k_2}-1}}\int_{p_{S1}}^{p_{S2}}p_S^{\frac{1}{k_S}+\frac{1}{k_2}-2}dp_S\right\} \tag{5.29}$$

$$= \frac{\alpha\, M_W\, p_{S2}}{R T_{S2}}\left\{(\beta+1)\left(1-\left(\frac{p_{S1}}{p_{S2}}\right)^{1/k_S}\right)-\left(\frac{p_1}{p_2}\frac{p_b}{\varphi p_{S2}}\right)^{\frac{k_2-1}{k_2}}\frac{1}{k_S\left(\frac{1}{k_S}+\frac{1}{k_2}-1\right)}\left(1-\left(\frac{p_{S1}}{p_{S2}}\right)^{\frac{1}{k_S}+\frac{1}{k_2}-1}\right)\right\} \tag{5.30}$$

Case 3: Variable Cavern Pressure and Constant Turbine Inlet Pressure

In the third case we consider, the air recovered from storage is throttled from the reservoir pressure p_s to the hp turbine inlet pressure p_1 such that the mass flow and expansion work output are constant in time. As in case 1, the integral representing the mechanical work in turbine expansion can be reduced to a simple time average, but in this case, the net air mass withdrawn from storage is a function of the storage pressure fluctuation over the range p_{S2} to p_{S1}:

$$\dot{m}_T = \frac{\Delta m_A}{t}\left(1 + \frac{\dot{m}_F}{\dot{m}_A}\right) \tag{5.31}$$

$$\Delta m_A = \frac{V_S p_{S2}}{R T_{S2}} - \frac{V_S p_{S1}}{R T_{S1}} = \frac{V_S p_{S2}}{R T_{S2}}\left(1 - \left[\frac{p_{S1}}{p_{S2}}\right]^{\frac{1}{k_S}}\right) \tag{5.32}$$

Substituting these into Equation (5.30) yields

$$\frac{E_{GEN}}{V_S} = \frac{\alpha\, M_W\, p_{S2}}{R T_{S2}}\left(\beta + 1 - \left(\frac{p_b}{p_2}\right)^{\frac{k_2-1}{k_2}}\right)\left(1 - \left[\frac{p_{S1}}{p_{S2}}\right]^{\frac{1}{k_S}}\right) \tag{5.33}$$

6

Battery Energy Storage

Isaac Scott and Se-Hee Lee

CONTENTS

Introduction

Energy based on electricity generated from renewable sources such as sun and wind offers enormous potential for meeting future energy demands. However, the use of electricity generated from these intermittent, renewable sources requires efficient electrical energy storage (EES). Efficient and durable electrical energy storage is one of the major limiting factors for widespread adoption of renewable energy. This is true from the level of the national and regional electricity grids down to the level of the home and automobile. Thus, for large-scale solar- or wind-based electrical generation to be practical, the development of new EES systems is critical for meeting continuous energy demands and effectively leveling the cyclic nature of these energy sources.

Chemical energy storage devices (batteries) and electrochemical capacitors (ECs) are among the leading EES technologies today.[1] Both are based on electrochemistry, and the fundamental difference between them is that batteries store energy in chemical reactants capable of generating charges, whereas electrochemical capacitors store energy directly as charges. Although the electrochemical capacitor is a promising technology for electrical energy storage—especially considering its high power capability—its energy density is too low to be considered for large scale energy storage. For this reason, electrochemical capacitors will not be covered in this chapter. The discussion in this chapter focuses on rechargeable battery technologies for large-scale energy storage.[2] Among various rechargeable battery technologies, four specific chemistries (lead–acid, sodium–sulfur (NaS), vanadium redox, and lithium ion (Li-ion) are reviewed in detail.[3] One case study based on the NaS system is discussed to show practical applications of batteries for large-scale energy storage applications.[4] Vanadium redox batteries (VRBs) are also reviewed because they can scale up to much larger storage capacities and show great potential for longer lifetimes and lower per-cycle costs than conventional batteries requiring refurbishment of electrodes. Finally lithium ion batteries are reviewed because they display very high potential for large-scale energy storage.

A battery contains one or more electrochemical cells; these may be connected in series or arranged in parallel to provide the desired voltage and power. The anode is the electronegative electrode from which electrons are generated to do external work. The cathode is the electropositive electrode to which positive ions migrate inside the cell and electrons migrate through the external electrical circuit. The electrolyte allows the flow of ions, for example, lithium ions in Li-ion batteries allow flow from one electrode to another. The flow is restricted to electrons and not ions. The electrolyte is commonly a liquid solution containing a salt dissolved in a solvent. The electrolyte must be stable in the presence of both electrodes.

The current collectors allow the transport of electrons to and from the electrodes. They are typically metals and must not react with the electrode or electrolyte materials. The cell voltage is determined by the energy of the chemical reaction occurring in a cell. The anode and cathode are, in practice, complex composites. They contain, besides the active material, polymeric binders to hold together the powder structure and conductive diluents such as carbon black to give the whole structure electronic conductivity so that electrons can be transported to the active material. In addition, these components are combined to ensure sufficient porosity to allow the liquid electrolyte to penetrate the powder structure and permit the ions to reach the reacting sites.

Secondary or Rechargeable Batteries

Secondary or rechargeable batteries are widely used in many applications such as starting, lighting, and ignition (SLI) automotive applications;

industrial truck materials handling equipment; emergency and standby power; portable devices such as tools, toys, lighting; and more significantly, consumer electronic devices (computers, camcorders, cellular phones). More recently, secondary batteries have received renewed interest as power sources for electric and hybrid electric vehicles and large-scale energy storage in a load leveling mode. Major development programs have been initiated to improve the performance of existing battery systems and developing new systems to meet the stringent specifications of these new applications.

The use of battery energy storage in utility applications allows the efficient use of inexpensive base load energy to provide benefits from peak shaving and handle many other applications. This reduces utility costs and permits compliance with environmental regulations. Analyses have determined that battery energy storage can benefit all sectors of modern utilities: generation, transmission, distribution, and end use. The use of battery systems for generation load leveling alone cannot justify the cost of the system. However, when a single battery system is used for multiple, compatible applications, such as frequency regulation and spinning reserve, the system economics are often predicted to be favorable.

Commercially available lead–acid batteries can satisfy the requirements for certain utility energy storage applications and are currently in use in several demonstration projects worldwide. Advanced batteries offer still greater potential for reduced costs and could enable market opportunities to be enhanced. These opportunities result from the predicted advantages of advanced batteries for lower cost, smaller system footprint, no maintenance, and high reliability even when used with highly variable duty cycles. Battery storage provides significant benefits in solar, wind, and other renewable generation systems where energy sources are intermittent. The batteries are charged when the source generates energy and the energy can then be discharged when the source is not available. Operating characteristics vary widely, depending on application. For photovoltaic systems, typical applications include village power, telemetry, telecommunications, powering remote homes, and lighting. The major candidates for electric vehicles (EVs), hybrid electric vehicles (HEVs), and electric utility applications in the near term are the rechargeable battery technologies now available commercially. Many of these have been improved over the last decade to meet the needs of the emerging applications. Further improvement may, in most cases, be necessary to effect economic viability.

Energy and Power

A quantitative comparison of four secondary battery systems is presented in Table 6.1.

TABLE 6.1

Characteristics of Major Secondary Systems

	Lead–Acid	NaS	Li Ion	Vanadium Redox
Chemistry:				
Anode	Pb	Na	C	$V^{2+} \leftrightarrow V^{3+}$
Cathode	PbO_2	S	$LiCoO_2$	$V^{4+} \leftrightarrow V^{5+}$
Electrolyte	H_2SO_4	β-alumina	Organic solvent	H_2SO_4
Cell voltage:				
Open circuit	2.1	2.1	4.1	1.2
Operating	2.0 to 1.8	2.0 to 1.8	4.0 to 3.0	
Specific energy and energy density:				
Wh/kg	10 to 35	133 to 202	150	20 to 30
Wh/L	50 to 90	285 to 345	400	30
Discharge profile	Flat	Flat	Sloping	Flat
Specific power (W/kg)	Moderate 35 to 50	High 36 to 60	Moderate 80 to 130	High 110
Cycle life (cycles)	200 to 700	2,500 to 4,500	1,000	12,000
Advantages	Low cost, good high rate	Potential low cost, high cycle life, high energy, good power density, high efficiency	High specific energy and energy density, low self discharge, long cycle life	High energy, efficiency, and charge rate, low replacement cost
Limitations	Limited energy density, hydrogen evolution	Thermal management, safety, seal and freeze–thaw durabilities	Lower rate (compared to aqueous systems)	Cross mixing of electrolytes

Lead–Acid Batteries

All lead–acid designs share the same basic chemistry. The positive electrode is composed of lead dioxide (PbO_2) and the negative electrode is composed of metallic lead (Pb). The active material in both electrodes is highly porous to maximize surface area. The electrolyte is a sulfuric acid, usually around 37% by weight when the battery is fully charged. The major starting material is highly purified lead. The lead is used for the production of alloys (for subsequent conversion to grids) and to produce lead oxides (for subsequent conversion first to paste and ultimately to the positive lead dioxide active material and the negative sponge lead active material).

The preparation of the active material precursor consists of a series of mixing and curing operations using a mixture of lead and lead oxide (PbO + Pb), sulfuric acid, and water. The ratios of the reactants and curing conditions (temperature, humidity, and time) affect the development of crystallinity and pore structure. The cured plate consists of lead sulfate, lead oxide, and some residual lead (<5%). The positive active material formed electrochemically from the cured plate is a major factor influencing the performance and life of a lead–acid battery. In general, the negative or lead electrode controls cold-temperature performance (such as engine starting).

Pure lead is generally too soft to serve as a grid material and was hardened traditionally by the addition of antimony metal. The amount of antimony varied between 5 and 12% by weight, generally dependent on the availability and cost of antimony. Typical modern alloys, especially for deep-cycling applications, contain 4 to 6% antimony. The trend in grid alloys is to use even lower antimony content, in the range of 1.5 to 2%, to reduce the required maintenance (water addition) of the battery. As the antimony content goes below 4%, the addition of small amounts of other elements is necessary to prevent grid fabrication defects and grid brittleness. These elements, such as sulfur, copper, arsenic, selenium, tellurium, and various combinations of these elements act as grain refiners to decrease the lead grain size.

Lead oxide is converted to a plastic dough-like material so that it can be affixed to the grids. Lead oxide is combined with water and sulfuric acid in a mechanical mixer. Sulfuric acid acts as a bulking agent—the more acid used, the lower the plate density will be. The total amount of liquid and the type of mixer used will affect final paste consistency (viscosity). The process is a form of extrusion by which the paste is integrated with the grid to produce a battery plate; it is known as pasting. The paste is pressed by hand trowel or by machine into the grid interstices. A curing process is used to form the paste into a cohesive, porous mass and to help produce a bond between the paste and the grid. The simplest cell consists of one negative electrode, one positive electrode, and one separator between them. Most practical cells contain 3 to 30 plates with the required number of separators. Individual or leaf separators are generally used. The use of "envelope" separators surrounding the positive or negative plate or both is becoming more popular in small, sealed cells, motive power, and standby batteries to facilitate production and control lead contamination during manufacture. Separators are used to electrically insulate each plate from its nearest counter-electrode neighbors but must be porous enough to allow acid transport into or from the plates. As a cell discharges, both electrodes are converted to lead sulfate. The process reverses on charge. The half cell reactions in discharge are as follows:

Positive: $PbO_2 + 3H^+ + HSO_4^- + 2e^- \rightarrow PbSO_4 + 2H_2O$

Negative: $Pb + HSO_4^- \rightarrow PbSO_4 + H^+ + 2e^-$

The overall discharge reaction is

$$PbO_2 + Pb + 2H_2SO_4 \rightarrow 2PbSO_4 + 2H_2O$$

As shown, the basic electrode processes in the positive and negative electrodes involve a dissolution–precipitation mechanism and not a solid-state ion transport or film formation mechanism. As the sulfuric acid in the electrolyte is consumed during discharge and produces water, the electrolyte is an "active" material and in certain battery designs can serve as the capacity-limiting material. As the cell approaches full charge and most of the $PbSO_4$ has been converted to Pb or PbO_2, the cell voltage on charge becomes greater than the gassing voltage (about 2.39 V per cell) and the overcharge reactions begin, resulting in the production of hydrogen and oxygen (gassing) and the resultant loss of water. In sealed lead–acid cells, this reaction is controlled to minimize hydrogen evolution and the loss of water by recombination of the evolved oxygen with the negative plate.

Stationary batteries are generally used to provide direct current (dc) power for controls and switching operations, as well as standby emergency power, in utility substations, power generation plants, and telecommunications systems. For the most part, these batteries operate under what is known as float charge. A charger keeps them at the full charge voltage with a small charging current, so that they are ready for use when needed. A battery experiences occasional discharges when a relay, breaker, or motor is energized and during outages. In this application, energy and power density are of secondary importance to long life and low maintenance. Partly for these reasons, stationary batteries have seen comparatively little development since their introduction in the early twentieth-century. The construction of these batteries tends to be very conservative. The care used in construction is reflected in their extremely long service lives, often extending to 30 to 40 years. The water lost to electrolysis during long periods of float charge must be replaced regularly. The batteries contain large electrolyte reservoirs with a quantity of excess electrolyte to extend the intervals between such maintenance operations.

Sodium–Sulfur (NaS) Batteries

Ford Motor Company is credited with the first research into the potential of the sodium–sulfur battery based on a β-alumina (β-Al_2O_3) solid electrolyte in the 1960s.[5] The basic cell structure and associated electrochemistry of the NaS cell are depicted in Figure 6.1. Liquid sodium is the active material in the negative electrode and the ceramic β-Al_2O_3 functions as the electrolyte. The cell is made in a tall cylindrical configuration, enclosed entirely by an inert metal container, and sealed at the top with an air-tight alumina lid. Such cells become more economical with increasing size. In commercial applications, the cells are arranged in blocks for better conservation of heat.

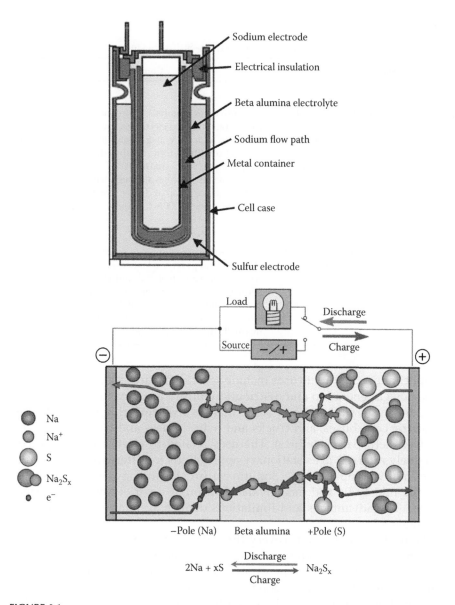

FIGURE 6.1
(a) NaS battery cell. (b) Electron and ion motion during discharge and charge cycles. (Courtesy of NGK Insulators Ltd.)

As the cells run, the heat produced by charging and discharging cycles is enough to maintain operating temperatures and no external heat source is required.

Rechargeable high-temperature battery technologies that utilize metallic sodium offer attractive solutions for many large-scale energy-storage

TABLE 6.2

Properties and Primary Limitations of Sodium–Sulfur System

Characteristic	Comments
Advantages	
Potential low cost relative to other advanced batteries; high cycle life; high energy; good power density; flexible operation; high energy efficiency; insensitivity to ambient conditions; state-of-charge identification	Inexpensive raw materials, sealed, no-maintenance configuration; liquid electrodes; low-density active materials; high cell voltage; cells functional over wide range of conditions (rate, depth of discharge, temperature); > 80% due to 100% Coulombic efficiency; reasonable resistance; sealed high-temperature systems; high resistance at top of charge; straightforward current integration due to 100% Coulombic operation
Limitations	
Thermal management; safety; seal and freeze–thaw durabilities	Effective enclosure required to maintain energy efficiency and provide adequate stand time; reaction with molten active materials must be controlled; cell hermeticity required in corrosive environment due to use of ceramic electrolyte with limited fracture toughness that can be subjected to high levels of thermally driven mechanical stress

applications. Candidate uses include several that are associated with electric power generation and distribution (utility load leveling, power quality, and peak shaving) and some that involve powering motive devices (electric cars, buses, and trucks and hybrid buses and trucks) and space power (aerospace satellites). The uses related to electric power are collectively referred to as stationary applications to differentiate them from the motive applications. Sodium–sulfur technology was introduced in the mid-1970s and has advanced through a variety of designs since then.[6] The advantages and limitations of such systems are summarized in Table 6.2.

From the time of its discovery through the mid 1990s, NaS system was among the leading candidates believed to be capable of satisfying the needs of a number of emerging energy storage applications that had very promising markets. The one application, however, that generated the most interest centered on powering EVs. Based on the large size of the potential market in combination with the inherent environmental advantages, many private companies and government organizations invested heavily in technology development. Significant advancements were made and because of acceptable performance, durability, safety, and manufacturability, at least four automated pilot-production facilities were built and operated by the mid 1990s. However, during this same period, government and industry realized that the public (especially in the United States) would not purchase pure battery-powered electric vehicles in high volumes. Their shorter ranges, lower power, and especially their higher cost compared with conventional internal combustion-powered vehicles were believed to ultimately make them unacceptable.

The basic cell structure and associated electrochemistry of the NaS technologies are depicted in Figure 6.1. NaS cells use solid sodium ion-conducting β-Al$_2$O$_3$ electrolytes. Cells must be operated at a sufficiently high temperature (270 to 350°C) to keep all the active electrode materials in a molten state and ensure adequate ionic conductivity through the β-Al$_2$O$_3$ electrolyte. During discharge, the sodium (negative electrode) is oxidized at the sodium β-Al$_2$O$_3$ interface, forming Na$^+$ ions that migrate through the electrolyte and combine with the sulfur being reduced in the positive electrode compartment to form sodium pentasulfide (Na$_2$S$_5$). The Na$_2$S$_5$ is immiscible with the remaining sulfur, thus forming a two-phase liquid mixture. After all the free sulfur phase is consumed, the Na$_2$S$_5$ is progressively converted into single-phase sodium polysulfides with progressively higher sulfur content (Na$_2$S$_{5-x}$). During charge, these chemical reactions are reversed. The half cell reactions in discharge are as follows:

$$\text{Positive:} \quad xS + 2e^- \rightarrow S_x^{-2}$$
$$\text{Negative:} \quad 2Na \rightarrow 2Na^+ + 2e^-$$

The overall discharge reaction is

$$2Na + xS \rightarrow Na_2S_x \ (x = 5 - 3) \ E_{OCV} = 2.076 - 1.78 \text{ V}$$

Although the actual electrical characteristics of sodium–sulfur cells are design-dependent, the general voltage behavior follows that predicted by thermodynamics. A typical cell response is shown in Figure 6.2. The figure is a plot of the equilibrium potential or open-circuit voltage and the working voltages (charge and discharge) as a function of the depth of discharge. The open-circuit voltage is constant (at 2.076 V) during the 60 to 75% of the discharge when the two-phase mixture of sulfur and Na$_2$S$_5$ is present. The voltage then linearly decreases in the single-phase Na$_2$S$_x$ region to the selected end-of-discharge point.

End of discharge is normally defined at open-circuit voltages of 1.78 to 1.9 V. The approximate sodium polysulfide composition corresponding to 1.9 V per cell is Na$_2$S$_4$; for 1.78 V per cell, it is Na$_2$S$_3$. Many developers choose to limit the discharge to less than 100% of theoretical (e.g., 1.9 V) for two reasons: (1) the corrosivity of Na$_2$S$_x$ increases as x decreases and (2) to prevent local cell over-discharge due to possible non-uniformities within the battery

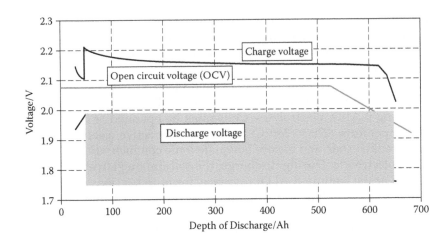

FIGURE 6.2
NaS cell voltage characteristics. (Courtesy of NGK Insulators Ltd.)

(temperature or depth of discharge). If the discharge continues past Na_2S_3, another two-phase mixture again forms, but now the second phase is solid Na_2S_2. The formation of Na_2S_2 in a cell is undesirable because high internal resistance, very poor rechargeability, and structural damage to the electrolyte can result.

Several other important characteristics of the sodium–sulfur electrochemical couple are evident from Figure 6.2. At high states of charge, the working voltage during charge increases dramatically due to the insulating nature of pure sulfur (shown also by the higher cell resistance). This same factor also causes a slight decrease in cell voltage at the start of discharge. At the C/3 discharge rate, the average cell working voltage is approximately 1.9 V. The theoretical specific energy of the electrochemical couple is 755 Wh/kg (to 1.76 V open circuit). Although not all of the sodium is recovered during the initial charge, cells subsequently deliver 85 to 90% of their theoretical ampere-hour capacity. Finally, the existence of wholly molten reactants and products eliminates the classical morphology-based electrode aging mechanisms, thus yielding an intrinsically high cycle life.

Compared to vanadium redox batteries (VRBs), NaS batteries have the advantage of extremely quick response time, making them more suitable for power quality applications (smoothing short term spikes in demand). It is believed that these advantages along with better round-trip efficiency are the reasons NaS batteries currently seem to have an edge with utilities seeking to delay transmission and distribution upgrades. American Electric Power (AEP) recently installed a $2.5 million, 7.2 MWh battery as described in Case Study 1.

Case Study 1: American Electric Power NaS Battery Project

Introduction

Rechargeable batteries will play an important role in delivering clean, reliable electricity to businesses and consumers. As recent power outages have demonstrated, battery technology is crucial for preventing the tremendous losses associated with momentary or prolonged power failures. To this end the sodium–sulfur battery has become one of the most promising candidates for commercial-scale stationary energy storage. In this case study, we discuss the installation of the first distributed energy storage system in the United States, describe the basic features of the sodium–sulfur battery, and summarize the rationale for its use.

Electrical energy comprises 12% of the total energy processed worldwide. The production of electricity is highly centralized and production usually takes place a long distance away from the end users. This delocalized electricity production increases the difficulty of stabilizing the power network, mainly due to a supply-and-demand imbalance. It is important to note that because the production of electricity is centralized, a complex system of energy production and transmission that makes little to no use of energy storage has been developed. The traditional method of dealing with increased electric demand has historically been to build more power plants and transmission lines, but increasingly dense urban areas and environmental and legislative restrictions make this cost prohibitive and outright impossible in some cases.

Storage capability is one of the main advantages of the NaS battery. Although a battery storage facility costs as much as a coal power plant that can supply the same amount of power, it has become impossible to build power plants in cities—precisely where the power is needed most.[7] From an efficiency point of view, the ultimate goal would be to generate the energy, transmit it, convert it, and then store it near the location where it is needed. Past battery technologies lacked the ability to address the changing demands for power back-up and storage and required bulky, costly equipment. Recent strides in technology have now made it possible to implement the first generation of large-scale energy storage devices. To this end, American Electric Power (AEP) decided to install the first commercial-sized NaS battery-based energy storage system in the United States at a chemical station in North Charleston, West Virginia.

Because the system was the first of its type, AEP received partial funding from the Energy Storage Program of the U.S. Department of Energy and Sandia National Laboratories. Additional incentives came from the manufacturer, NGK Insulators Ltd., in the form of preferential prices for future purchases of NaS batteries for a limited number of years. Installation of the 1.2-MW NaS-based, distributed energy storage system (DESS) was completed in 9 months and entered commercial operation on June 26, 2006.

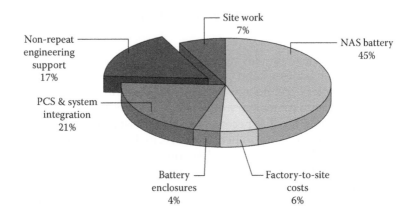

FIGURE 6.3
Major cost components for installed NaS-based DESS. (Courtesy of NGK Insulators Ltd.)

Figure 6.3 provides a breakdown of the associated costs for the installation of the DESS. The other costs included the NaS batteries purchased from NGK (Nagoya, Japan), the power conversion system (PCS), built and installed by S&C Electric Company, and the 17-foot high steel enclosures that hold 20 NaS batteries. As is the case with initial implementations of new technologies, the cost of this DESS installation was affected by some factors unique to first installations and represent costs not expected in future installations.

Rationale for Using NaS-Based DESS

Any company must ask several questions when considering the purchase of a new capital asset, especially one involving new technology: (1) Is the purchase warranted? (2) What are the current limitations of our infrastructure and what future problems may we encounter? (3) What benefits will our company gain? (4) Are there cheaper alternatives? In addition to these issues, AEP is experiencing exponential growth in the number of customer-owned distributed generation (DG) systems that are requesting connections to the grid. These DG systems were made possible by the advances in cost and efficiency of many renewable energy systems (solar panels, wind turbines, etc.). Among the operational challenges that AEP faces are: (1) non-optimized location of generation, (2) uncertain availability of generation, (3) low reliability of available generation, (4) inadequate dispatching or scheduling control over generation, and (5) safety concerns related to energy backfeed. After evaluating the benefits, strengths, and weaknesses of various solutions, AEP concluded that energy storage would allow grid operators to maintain control over the grid, improve service reliability, and would accrue benefits from the presence of the customer-owned DGs on the grid. The most cost-effective product on the market was the NaS-based DESS.

Long-Term and Short-Term Benefits of Energy Storage

Based on widespread energy storage deployment on a grid, strategic decisions considering the long- and short-term impacts require review. The long-term benefits of well-penetrated distributed energy storage include:

- Improved system control and reliability to cope with adverse impacts of widespread and uncontrolled customer-owned DGs.
- Enhancement of DG penetration by reducing its required size.
- Improved system reliability due to intentional islanding that can negate brownouts and even blackouts.
- Base loaded power
- Improved asset management and extending useful equipment life by reducing peak loads at all system levels.
- Reducing equipment cost by lowering required power ratings.
- Opportunities to offer energy arbitrage in deregulated environments.
- Provision of voltage and frequency regulation benefits.

The most important short-term benefits of distributed energy storage for utilities involve "buying time" by

- Deferring upgrade capital expenditures by reduction of load peaks.
- Improving service reliability where conventional solutions (constructing new power lines or substations) may not be readily available or would take several years to implement.
- Allowing more time for service restoration during scheduled or accidental power interruptions due to ability to provide interim power to customers.

Load leveling is initially based on predictions of daily and seasonal needs when production is not sufficient. Traditional batteries are difficult to scale up, contain hazardous chemicals, and do not meet the performance and life standards required by distributed and stationary energy storage applications. Presently the NaS-based DESS is one of the most cost-effective systems on the market, but the utility industry continues to search for the next generation of battery systems that can provide reliable backup for businesses and consumers during power outages, decrease costs of electrical distribution and generation by shaving peak costs, and reduce plant costs by delivering cleaner electricity while retaining the ability to handle multiple levels of storage including peaking, intermediate, and base loads. The challenges for new battery technology are great, but with energy shortages looming, we will see more DESS facilities installed.

Case Study 2: Xcel Energy to Test Storage of Wind Power Using 1-MW Battery System

Xcel Energy soon will begin testing a cutting-edge technology to store wind energy in batteries. It will represent the first use of the technology in the United States for direct wind energy storage. Integrating variable wind and solar power production with the needs of a power grid is an ongoing concern for the utility industry. Xcel Energy will begin testing a 1-MW battery storage technology to determine its ability to store wind energy and move it to the electricity grid when needed. When fully charged, the battery could power 500 homes for over 7 hours. Xcel Energy signed a contract to purchase a battery from NGK Insulators Ltd. that will be an integral part of the project. NaS batteries are commercially available and versions of this technology are already in use in Japan and in a few United States applications, but Xcel's installation will be the first United States application of the battery as a direct wind energy storage device. The 20- to 50-kW battery modules will be roughly the size of two semi trailers and weigh approximately 80 tons. They will be able to store about 7.2 MWh of electricity, with a charge–discharge capacity of 1 MW. When the wind blows, the batteries are charged. When the wind calms, the batteries supplement the power flow. Partners in the project with Xcel Energy include the University of Minnesota, the National Renewable Energy Laboratory, the Great Plains Institute, and Minwind Energy LLC. Xcel is testing emerging technology and energy storage devices as part of its overall "smart grid" strategy intended to modernize and upgrade its grid to allow easier integration of renewable energy sources.

Vanadium Redox Batteries

The technical boundary conditions to electricity storage results from two distinctly different requirements based on the service conditions of the battery: (1) load leveling and peak shaving and (2) seasonal energy storage. For load leveling applications, the storage medium must have a high power density, be able to recharge at high rates for many cycles, and be able to withstand deep discharges. For seasonal storage, a battery must have large capacity, low self-discharge rate, and be capable of operating on a great number of shallow cycles. The current trend in wind energy is to install turbines in a large wind farm connected to a single point in the grid.

Due to wind power fluctuations and limited wind predictability, the ideal energy storage device for wind farms would achieve load leveling and stabilization of turbine output. The storage device would store power at night when the wind blows hardest and electricity usage and rates are low, then deliver power to the grid during the day when transmission lines reach capacity, and thus allow a utility to profit from higher daytime rates. To attain a higher degree of controllability, an energy storage system must have the capacity to store a lot of energy for short periods and deliver it during

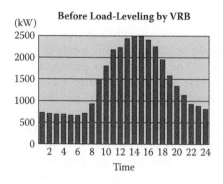

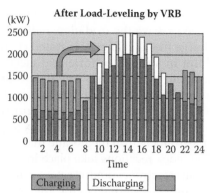

FIGURE 6.4

Load leveling applications of a vanadium redox flow battery (VRB) for Tomari Wind Hills power station of Hokkaido Electric Power Co., Inc. (Graph courtesy of Sumitomo Electric Industries Ltd., 2001.)

peak load hours. Figure 6.4 provides a graph of the load leveling capabilities of a vanadium redox flow battery for a wind turbine application.[8]

Properties of Other Electrochemical Storage Devices

Before we discuss vanadium redox flow technology, it is important to first explain and define fuel cells and flow batteries. Fuel cells convert the chemical energy available from the oxidation of a fuel (usually hydrogen) by an oxidant (oxygen from air, for example) directly into electricity. The operating behavior is similar to that of a primary battery with infinite capacity. A flow battery is a type of rechargeable secondary battery in which energy is stored chemically in liquid electrolytes. The electrolytes contain dissolved electro-active species that flow through a power cell that converts chemical energy to electricity.

Simply stated, flow batteries are fuel cells that can be recharged. From a practical view, storage of the reactants is very important. The storage of gaseous fuels in fuel cells requires large, high pressure tanks or cryogenic storage that is prone to thermal self-discharge. Vanadium redox batteries have the advantage of being able to store electrolyte solutions in non-pressurized vessels at room temperature. Vanadium based redox flow batteries hold great promise for storing electric energy on a large scale (theoretically, they have infinite capacity). They have many attractive features including independent sizing of power and energy capacity, long lifetimes, high efficiency, and fast response, and relatively low cost (small initial investment and reduced operational expenditures).

Vanadium Redox Flow Batteries

A redox flow cell is an electrochemical system that allows energy to be stored in two solutions containing different redox couples with electrochemical potentials sufficiently separated from each other to provide an electromotive force to drive the oxidation–reduction reactions needed to charge and discharge the cell.[9] Energy is stored chemically in different ionic forms of vanadium in a dilute sulfuric acid electrolyte. The electrolyte is pumped from separate storage tanks into flow cells across a proton exchange membrane where one form of the electrolyte is electrochemically oxidized and the other is electrochemically reduced (Figure 6.5).

The two electrolytes do not mix together; they are separated in the cells by an extremely thin membrane that allows only selected ions to flow through. The redox reactions take place in the cells on inert carbon felt polymer composite electrodes and create a current that becomes available to do work through an external circuit, after which charging the battery can reverse the reaction.

Before the VRB appeared, the main disadvantage to flow batteries was that the two liquid electrolytes were made of different substances and separated by a thin membrane that was eventually permeated, after which the two substances would mix and render the battery useless. The main advantage of the VRB system is that vanadium is present in both the positive and negative electrolytes, but in different oxidation states. Vanadium has four oxidation states: V^{+2}, V^{+3}, V^{+4}, and V^{+5}. The VRB thereby exploits the ability of vanadium to exist in solution in four different oxidation states—an ability shared only with uranium and other heavy radioactive elements.

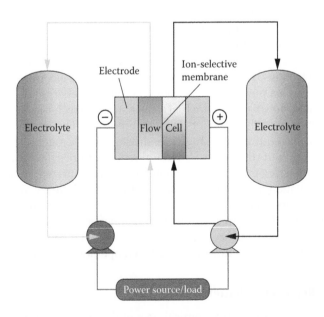

FIGURE 6.5
Flow battery that stores energy in liquid electrolytes. (Courtesy of VRB Power Systems Inc.)

The electrolyte is composed of vanadium salts dissolved in sulfuric acid, and if the electrolytes are accidentally mixed, the battery suffers no permanent damage. The standard cell potential E^0 is 1.26 V at concentrations of 1 M, but under actual cell conditions, the open circuit voltage (OCV) is 1.4 V at 50% state of charge (SOC) and 1.6 V at 100% SOC. The electrode reactions occur in solution. The reaction at the negative electrode in discharge is $V^{2+} \rightarrow V^{3+} + e$ and the reaction at the positive electrode is $V^{5+} + e \rightarrow V^{4+}$. Both reactions are reversible on the carbon felt electrodes. An ion-selective membrane is used to separate the electrolytes in the positive and negative compartments of the cells. Cross-mixing of the reactants would result in a permanent loss in energy storage capacity for the system because of the resulting dilution of the active materials. Migration of other ions (mainly H^+) to maintain electroneutrality, however, must be permitted. Thus, ion-selective membranes are required.

Because the electrolyte is returned to the same state at the end of every cycle, it may be reused indefinitely. The negative half-cell uses the V^{2+}–V^{3+} redox couple, whereas the positive half-cell uses the V^{4+}–V^{5+} redox couple. The positive and the negative vanadium redox couples show relatively fast kinetics that allow high Coulombic and voltage efficiencies to be achieved without costly catalysts. However only V^{5+}, V^{4+}, and V^{3+} are stable in air; V^{2+} is easily oxidized by atmospheric oxygen, which must be taken into account when servicing the negative electrolytic solution. However, different oxidation states are not sufficient to make an element work in a liquid electrolytic solution. The element must also be soluble. Although V^{2+}, V^{3+}, and V^{4+} species are easily soluble in sulfuric acid, the long-term stability of concentrated V^{5+} solutions is rather limited due to the formation of insoluble V_2O_5 precipitates at elevated electrolyte temperatures. It must be noted that 0.9 molar V^{5+} solutions are stable even at elevated temperatures, and increasing the concentration of sulfuric acid can increase stability of V^{5+} solutions.

As the reactions involve only dissolved salts, the electrode acts only as a site of reaction and does not participate in the chemical process. Therefore, it does not suffer ill effects due to changes in composition. Because it does not experience a physical or chemical change, it can carry out a large number of charge and discharge cycles without a significant decrease in capacity. The electrodes are located in the reaction cells that are grouped together in a series of blocks known as stacks. Each stack contains a series of conducting (bipolar) plates with a positive electrolyte on one side and a negative on the other. Because they share the same electrolyte, each cell of a flow battery is practically identical, unlike conventional batteries connected in series whose available power is limited by the poorest battery in the string. Thus the user must depend on the manufacturer to produce batteries with little variability.

The use of solutions to store energy means that system power and storage capacity are independent, making vanadium batteries scalable to a wide range of voltages, currents, and capacities and allowing them to be tailored

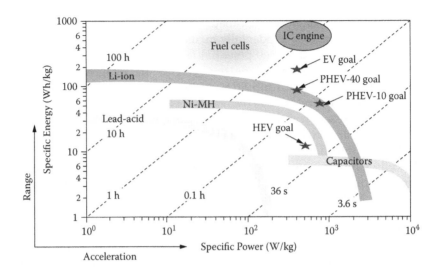

FIGURE 6.6
Plot of energy versus power for various energy systems. (From http://berc.lbl.gov.venkat.
Ragone-consruction.pps)

for diverse applications. This means that a VRB system can produce more kilowatts of power simply by resizing the stacks and/or supply more energy by increasing the size of the electrolyte storage tanks. The system theoretically has no limit to the amount of energy it can provide.

One may therefore ask, "Given all these advantages, why don't we see more VRBs in operation?" The answer mainly involves energy density, size, cost, possible irreversible precipitation of V_2O_5, and the infancy of the technology that makes most companies hesitant to invest in technology that has not been fully tested. The energy density of a VRB is limited by V_2O_5 and has been measured to be ~167 Wh/kg. Figure 6.6 provides a comparison of energy densities of other systems. For example, a 600-MWh vanadium redox flow battery system would require 30 million liters of electrolyte. If stored in 6-m high tanks, its footprint would be the size of a football field. In general, the following conditions must be met for the operation of vanadium redox flow cells:

- Electrodes require high electric conductivity and good wettability.
- Charging voltage must be limited to a maximum of 1.7 V to avoid damage to the carbon current collectors.
- Good electrical contact to the bipolar plates and current collectors is essential and best achieved when the activation layers are thermally bonded to the current collector.
- Access of oxygen to the negative electrolyte compartment must be avoided.

Note that high Coulombic efficiency is observed at high flow velocities; therefore higher current densities may be achieved when operating at these velocities.

Commercial Application: Cellstrom

Now that the vanadium redox flow battery system has been explained, we can discuss a current commercial application of this technology. Cellstrom, a German company, has developed a complete energy storage system (ESS) utilizing VRB technology.[6] The FB 10/100 consists of a vanadium redox flow battery with a smart controller and configurable power electronics housed in a weather-proof container. The system can be charged and discharged up to 10kW and provides 100kWh of energy (10 h discharge at full power). The battery can be connected to photovoltaic devices, wind turbines, diesel, petrol, gas, and biogas generators, fuel cells, and water turbines to form discrete autonomous power suppliers or can work as part of a micro-, mini- or smart grid. The FB 10/100 is divided into fluid, electrical, thermal, and safety systems.

The fluid system is seen in Figure 6.7, which also provides a layout diagram. Two tanks that house 2500 liters of electrolyte each are located on the lower level. Chemically resistant pumps provide the pressure to move the electrolyte into the stacks above. Even if the electrolytes mix, a smart controller automatically compensates for this by opening a rebalance valve to maintain the desired potential across the electrodes.

The electrical system is configured to meet the demands of the customer. The stacks are charged via a terminal connection to an outside power source. During discharge, the dc power of the batteries is converted to ac via inverters on the opposite sides of the stacks. An interface cabinet (Figure 6.8) provides lightning protection, AC fuses, and connection points for the load.

The operating temperature of the system ranges from 5 to 40°C. The temperature is kept within these parameters via the smart controller. This control system uses ventilation fans to cool the apparatus using outside air when the temperature exceeds operational temperature. In addition, the electrolyte solution acts as a coolant as it is pumped through the stacks to allow better heat exchange and helps to reduce thermal management problems.

When the temperature falls below the operational level, the smart controller seals the container. The heat generated from the system is then allowed to keep the cell operating within its temperature range. The temperature of the electronics side of the FB10/100 is also monitored and regulated by a separate ventilation system. Thermal separation of the fluid and electrical components also helps prevent local hotspots.

Finally, safety is built into the unit to protect against lightning, fire, spillage, and hydrogen production. The battery and electronics are housed in a sturdy container fitted with lightning protection for outdoor mounting. Any electrolyte that spills due to damaged stacks or lines returns to the tank via a

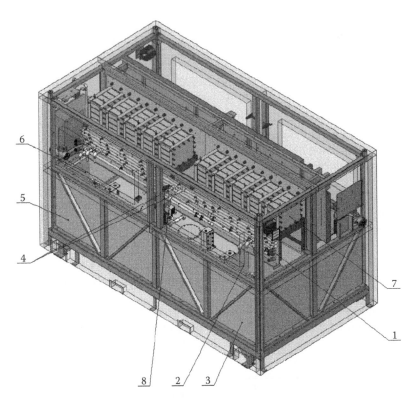

FIGURE 6.7
Fluid thermal safety system of Cellstrom's FB10/100. 1. Fluid lines. 2. Positive electrolyte pumps. 3. Positive electrolyte tank. 4. Return lines. 5. Negative electrolyte tank. 6. Negative electrolyte pumps. 7. Stacks (also called cell stacks or modules). 8. Rebalance valve. (From html://www.cellstrom.com)

drainage system. Leakage from the main tank is minimized by a secondary containment device containing leakage sensors. In addition, because the lines are kept at pressures of 1 bar above atmosphere, leakage in the lines will not cause significant spraying. As with all batteries containing water, hydrogen evolution must be monitored to prevent potential explosions. Fortunately the amount of hydrogen produced by this system is very small and is easily collected and vented from the tanks.

The FB10/100 is an environmentally friendly system because it contains no heavy metals. If the electrolyte is kept free of contamination, it can be reused indefinitely. In addition, 99.9% of the plastics used are non-halogenated and individual components can be replaced without the need to discard large components.

Finally the system was designed to require minimum maintenance. The FB10/100 has a design life of 20 years. While some of the stacks and pumps may have to be replaced during that lifetime, the fluid and electrical systems

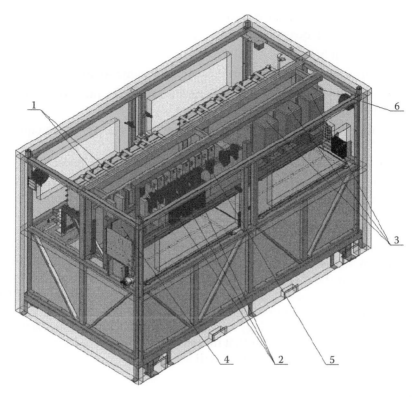

FIGURE 6.8
Electrical system of Cellstrom's FB10/100. 1. Stack terminal connectors. 2. Charge. 3. Inverter.
4. Smart controller. 5. Fuses. 6. Interface cabinet. (From html://www.cellstrom.com)

are separated to ensure contamination remains minimal if the need arises
to service only one part of the system. The smart controller monitors the
battery state and electrical system, and with a built-in wireless modem it
can transmit data and servicing messages to any computer or hand-held
device.

To summarize our discussion of the vanadium redox flow battery, its main
advantages are

- Storage of energy in separate tanks away from the cell stack
- Capability of increasing energy capacity simply by adding more
  solution
- Ability of electrolyte solution to act as a coolant when pumped
  through the stacks
- Lack of contamination from cross mixing of electrolytes
- Low cost based on indefinite lifetimes of solutions

- High-energy efficiencies arising from electrochemical reversibility of vanadium redox couples
- Recharge at high rates
- Uniform cell potential achieved by reuse of same solution
- Simple monitoring and maintenance; no need for monitoring and adjustment of individual cells
- Monitoring electrolyte state of charge by using the Nernstian equation $E = E^0 + \dfrac{RT}{nF} \ln \dfrac{a_O^{v_O}}{a_R^{v_R}}$ to measure the capacity of the battery.
- No need for overcharging for cell equalization; hydrogen explosion hazard is eliminated

The vanadium flow technology is still in its infancy. For it to become a more viable power storage solution, three requirements must be met: (1) developing a way to dissolve more V^{5+} into the electrolyte solution; (2) finding a replacement for the membrane—the flow battery's most expensive component; (3) achieving greater electrolytic energy density. Stabilizing a future national grid that draws most of its power from renewable sources may seem like a tall order for a technology that delivers megawatts, not gigawatts, of power, but one must remember that this technology is still in its infancy and as it matures many of its limitations will be solved.

Lithium Ion Batteries

Until the lithium ion (Li-ion) battery was introduced in the early 1990s nickel–metal hydride batteries were the industry standards for portable electronic devices. Since their introduction, Li-ion batteries have made significant strides in weight, capacity, and power, compared to their nickel–metal hydride competitors. Li-ion batteries are used extensively in consumer electronics, but their use in larger scale applications such as electric vehicles has been limited to date. The primary reasons for the delayed deployment are safety and high cost; and both issues are related. Although costs can be reduced with economies of scale of larger batteries, larger batteries are more difficult to cool and thus more prone to heat gain and the effects of high temperatures.

Ten years before the introduction of Li-ion batteries, the development of new hydride materials for nickel–metal batteries achieved a 30 to 40% increase in energy density compared to their nickel–cadmium predecessors. This was a quantum leap in battery technology and served to make nickel–metal-hydride the leader for use in portable electronics.

However, a number of issues surround nickel–metal hydride batteries: limited discharge currents, high levels self-discharging, long recharge times, and heating during recharging. The introduction of Li-ion batteries seemed to solve a number of issues inherent in nickel-based technology.[10] Li-ion batteries offered an even greater energy density then nickel–metal hydride batteries, improved self-discharge, shorter charge times, and an increased discharge voltage of about 3.7 V compared to 1.4 V for nickel–metal hydride batteries.

Li-ion batteries look similar to most batteries. Inside a cylindrical metal casing are three sheets tightly spiraled together and surrounded by a liquid electrolyte. The three sheets serve as a positive electrode, a negative electrode, and a separator. The positive electrode is generally lithium cobalt oxide ($LiCoO_2$) and the negative electrode is generally graphite (C_6). During charging, Li ions flow from the positive electrode ($LiCoO_2$) through the porous separator and intercalate into the negative electrode (C_6). During discharging, the ions flow in the opposite direction and reintercalate into the $LiCoO_2$. When the Li ions detach and move to the opposite sheet, an electron will be detached as well because Li ions are positively charged. This electron will flow out of the battery and to the component it is running and then back in to the opposite side to again meet up with the Li ion. This process can be seen in Figure 6.9.

Although Li-ion batteries appear effective for improving the limits of batteries, they are not perfect.[11–13] For example, all Li-ion batteries require complex circuitry to prevent overcharging. Overcharging can compromise the stability of the cathode, resulting in the breakdown of a battery. Complex circuitry includes a shutdown separator for over-temperature, tear-away tab for internal pressure, vent for pressure relief, and thermal interrupt for over-current and overcharging. Another major restriction is the cost of production. On average, a Li-ion battery costs 40% more to manufacture than a nickel–cadmium battery. The high price may be attributed to

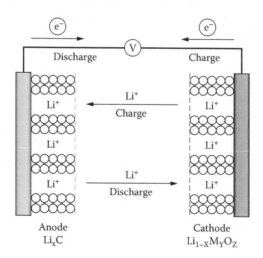

FIGURE 6.9
Charge and discharge cycles of typical lithium ion battery.

the high cost of cobalt and to the complex protective circuitry required by every battery.

Although cost and overcharging are issues for Li-ion batteries, the most prevalent problem seems to be overheating, especially in larger applications.[14] Overheating can cause significant decreases in performance ratings and even catastrophic failure. If Li-ion batteries are to be used more efficiently and for larger scale applications, the heat issues must be resolved. The primary concerns regarding high temperatures in Li-ion batteries are (1) increase in the probability of thermal runaway, (2) loss in capacity, (3) increased undesirable side reactions.

Thermal Runaway

The potential for thermal runaway is the most important concern with high-temperature operation. If a battery consistently generates more heat than it can dissipate, a condition known as thermal runaway can occur.[15,16] It can eventually lead to leaks, venting of gas, and possible explosion or fire. Recent recalls of laptop batteries due to isolated thermal runaway events increased this concern. To avoid thermal runaway, temperatures in a cell must be kept below 105 to 145°C, depending on the state of charge (SOC).

Thermal runaway, or the extreme overheating of a battery, can arise for a number of reasons. Safety issues generally occur if a cell exceeds the critical temperature above which the increase of temperature is irreversible due to the heat produced by the cathode, anode, and electrolyte and their interactions above the critical temperature.

Another cause of thermal runaway may be excessive ambient temperatures. Computers, cell phones, and other portable devices are often exposed to high ambient temperatures that can be caused by nothing more than the sun. Portable electronic devices stored in a car on a hot day or left next to a heater will be exposed to temperatures well above the level necessary to trigger thermal runaway. Excessive ambient temperatures will lead to electrolyte heating, thus increasing the exothermic chemical reactions.

A third source may be overcharging, leading to build-up of lithium deposits that eventually penetrate the separating film and short the two electrodes. All Li-ion batteries are equipped with protective circuitry to avoid overcharging, but if the protective circuitry fails, Li ions will build up on the graphite anode, forming a lithium dendrite. If the charging continues, the dendrite will grow until it penetrates the separator, creating a short circuit by connecting with the cobalt oxide cathode.

Regardless of the cause of thermal runaway, the result is the same and in many cases very dangerous. If Li-ion batteries are to be used on a larger scale and work more safely, they must be unaffected by higher temperatures or cooled more effectively to maintain lower temperatures. Many theories for solving this issue have been proposed but they are too costly or significantly alter the weight, size, and durability advantages of Li-ion batteries.

Capacity Fading

Capacity fading is a problem at lower temperatures. Capacity fading is a reduction in battery capacity after repeated cycles. Li-ion batteries show better performance than other secondary batteries applicable to EVs, but still show capacity fading after a high number of cycles. Elevated temperatures accelerate the rate at which capacity fading occurs with most lithium intercalation compounds. The number of cycles when the batteries are operated at 50°C or above are well below the EV requirement of >5000 cycles.[17] It is known that most capacity loss at these high temperatures is due to degradation of the electrode materials, particularly after a great number of cycles. Capacity fading can also occur when a battery is stored at a high temperature. Storage at 60°C for 60 days resulted in a 21% reduction in Li-ion battery capacity.

Loss of High-Rate Discharge Capacity

Operating temperatures of batteries vary from about −40 to +150°C. Li-ion batteries typically fall in the range of 1 to 35°C. Unlike other types of batteries, Li-ion battery performance can be affected significantly by operating temperatures outside this range.[18] The increased temperature will cause an exponential increase in the rate at which undesirable chemical processes occur.

These chemical reactions cause higher internal impedance that shortens the life of a battery and in some cases leads to the decomposition of the battery. Even a slightly elevated operating temperature of 40°C can diminish battery performance up to 35%. The batteries of high-power EVs must be able to discharge at high rates. In some cases, the increase in storage temperature (even to 60°C) reduces the high-rate discharge capacity by over 90%. Due to the requirement of high-rate discharge for EVs, increased temperatures must be avoided.

To make Li-ion batteries the industry standard for large applications, overheating and thermal runaway are not the only issues to consider. Costly production is another issue. As stated earlier, the increased expense can be attributed mostly to the costly metal cobalt used in the cathodes. Despite its stability and high-rate capability, $LiCoO_2$ is plagued by toxicity and the high cost of Co. If Co use could be eliminated, the cost of production would drastically be reduced. In the past few years, a large research effort has focused on replacing $LiCoO_2$ cathodes. As a result, Co has been partially and fully replaced with Ni and/or Mn.[19] One viable $LiCoO_2$ replacement material is the $Li(Ni_{1/3}Mn_{1/3}Co_{1/3})O_2$ layered structure.

Other researchers believe they have found the cost solution in a cathode made from a lithium iron phosphate ($LiFePO_4$). This cathode involves inexpensive raw materials, but differs significantly from the lithium manganese titanium oxide cathodes now in production.[20] The lithium iron phosphate must undergo a much more rigorous and costly production.

Many companies have invested heavily in research to find new cathode materials that could lower the cost of Li-ion batteries. They have developed a number of probable solutions to lower the cost, including the two cathode materials cited above. If one of these solutions is implemented for large-scale production, it would significantly decrease the cost of producing Li-ion batteries and spur further research and development.

The current safety issues surrounding thermal runaway and the high costs of producing Li-ion batteries have prevented them from becoming the industry standards for both small and large applications.[21] However, recent research has proven that with relatively simple modifications for smaller production costs, it is possible to produce a safe and reliable large-scale Li-ion battery. If designers were to implement the heat-reducing mechanical and safety features of the battery pack along with cost reductions and power increases for cathode materials, the Li-ion battery could efficiently and cleanly power everything from cell phones and computers, to cars and solar and/or wind power plants into the next century.

References

1. Linden, D. and Reddy, T. B. 2002. *Handbook of Batteries*, 3rd Ed. McGraw Hill, New York.
2. Electric Power Research Institute and United States Department of Energy. 2003. *Handbook of Energy Storage for Transmission and Distribution Applications*. December.
3. www.mines-energie.org/Conferences
4. Nourai, A. 2007. Report: Installation of the First Distributed Energy Storage System (DES) at American Electric Power (AEP). Sandia National Laboratories, Albuquerque, NM, No. 3580.
5. Yoa, Y.F.T. and Kummer, J.T. 1967. Ion exchange properties and rates of ionic diffusion in beta alumina. *Journal of Inorganic Nuclear Chemistry*.
6. Cellstrom GmbH, Wiener Neudorf, Austria. FB10/100 Technical Description. www.cellstrom.com
7. Storage: the next generation. Why build a new power plant when the technology exists to store excess megawatts until needed? *Mugnatto-Hamu*, Adriana, April 9, 2006.
8. Joerissen, Garche, Fabjan, Tomazic. 2004. Possible use of vanadium redox flow batteries for energy storage in small grids and stand-alone photovoltaic systems. *Journal of Power Sources*, 127, 98.
9. Bindner, Ahm, Ibsen. 2007. Vanadium redox flow batteries: installation at Riso for characterization measurements. Wind Energy Department, Riso National Laboratory, DTU.
10. The element that could change the world. *Discovery Magazine*. discovery-magazine.com/2008/oc/29-the-element-that-could-change-the-world/article_print

11. Akimoto, J. November 3, 2008. Opening the way to a low-cost secondary lithium ion battery. *AIST*

12. http://www.aist.go.jp/aist_e/annual/2006/highlight_p13/highlight_p13.html

13. Balbuena, Perla, Yixuan, Wang. 2004. *Lithium Ion Batteries*. Imperial College Press.

14. Bullis, K. May 22, 2007. Lithium ion batteries that don't explode. *Technology Review*. http://www.technologyreview.com/Energy/18762/?a=f.

15. Durrant, M. November 3, 2008. Thermal management. *MPower*. http://www.mpoweruk.com/thermal.htm.

16. Buchmann, I. Is lithium ion the ideal battery? http://www.batteryuniversity.com/print-partone-5.htm

17. Berdichevsky, G., Kelty, K., Straubel, J.B. et al. The Tesla Roadster battery system. http://www.teslamotors.com/display_data/TeslaRoadsterBatterySystem.pdf

18. Jurkelo, I. Advantages and disadvantages of the nickel–metal hydride battery. http://e-articles.info/e/a/title/Advantages-and-disadvantages-of-the-Nickel-Metal-Hydride-(NiMH)-Battery/

19. Buchmann, I. How to prolong lithium-based batteries. http://www.batteryuniversity.com/parttwo-34.htm

20. Sivashanmugam, A. 2004. Performance of a magnesium–lithium alloy as an anode for magnesium batteries. *Journal of Applied Electrochemistry*, 34, 1.

21. Whittingham, S. BATT program SUNY Binghamton. http://berc.lbl.gov/BATT/BATT%20summaries%202006.pdf

7

Solar Thermal Energy Storage

Carl Begeal and Terese Decker

CONTENTS

Parabolic trough for concentrating solar energy to fluid in a central tube. (From http://www.
schottsolar.com/us/products/concentrated-solar-power/concentrated-solar-power-plants/)

Introduction to Thermal Energy Storage

A challenge in utilizing renewable energy sources, particularly solar and wind energy, is dealing with their intermittent availability due to weather, diurnal solar variation, and seasonal changes. One way the variable nature of solar resources can be effectively managed is with supplemental thermal energy storage (TES). Variable energy can be stored when excess power is available and then discharged during periods of limited or no solar production to meet demands for continuous power production and/or to satisfy peaks in demand.

Beyond creating high value dispatchable power at times of low insolation, TES has many benefits as compared to other storage types and especially as compared with a system with no energy storage. As shown in Figure 7.1, the round-trip efficiency of TES is very high and exceeds that of other storage options. TES offers high storage efficiency for multiple hours at modest cost. Additionally, it allows stored power to match peak demand periods, making solar thermal energy technologies competitive with gas turbines as capacity power sources. TES is a deployable, proven technology and can be implemented immediately without further research. These topics and others are discussed in detail in this chapter.

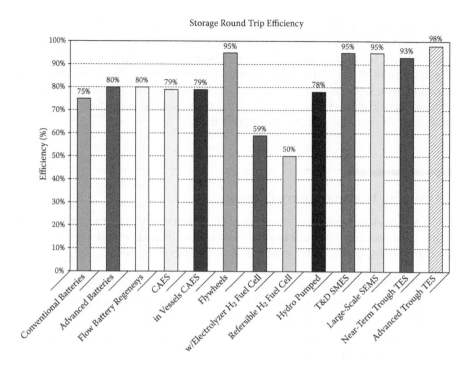

FIGURE 7.1
Comparison of efficiencies of several energy storage technologies. (From Turchi, C. July 21, 2008. Thermal Energy Storage for Concentrating Solar Power Plants. National Renewable Energy Laboratory.)

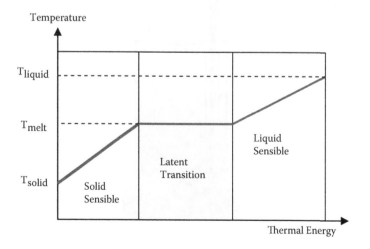

FIGURE 7.2
Plot of temperature versus thermal energy for material undergoing phase change.

TES is a reserve of energy in the form of heat that can be deployed independently of primary resources. It was developed to assist with several applications of energy production and heating, including supplementing the loading of utility scale solar thermal power plants. TES can be stored as sensible heat, latent heat, or the potential heat of recombination in reversible thermochemical reactions. This chapter addresses these modes of TES and discusses design instructions and applications.

Physics of Thermal Energy Storage

TES requires the increase or decrease of the internal energy of a substance by heating or cooling. Thermal energy is stored in materials that are classified by one of three methods by which they store energy as heat:

Sensible heat *involves a temperature change of the storage material.*

Latent heat *requires an isothermal phase change of the storage material (melting, freezing, vaporization, fusion, and crystallization).*

Heat of reaction or thermochemical heat *is the result of a reversible thermochemical reaction by the storage material.*

Figure 7.3 illustrates the temperature change when thermal energy is added to a nonreacting material. As heat is added to the material, its temperature

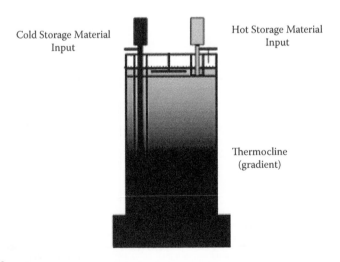

FIGURE 7.3
Thermocline technology schematic. (From National Renewable Energy Laboratory. 2008. TroughNet: Parabolic Trough Thermal Energy Storage Technology. http://www.nrel.gov/csp/troughnet/thermal_energy_storage.html)

increases until a phase change occurs. During the phase change, the solid bonds are broken at a constant temperature as heat is added to the material. This is the fundamental difference between sensible and latent heat. After the material has completed the phase change, in this case from a solid to a liquid, the material again undergoes an internal energy and temperature increase with the application of heat.

In general, it is desirable for TES to use materials that possess a large change in internal energy per unit volume and/or mass to minimize the space needed to store the desired amount of energy. To be economically competitive in commercial solar systems, it is important to employ substances with a high internal energy change per unit cost. Other properties such as vapor pressure, toxicity, and corrosiveness also must be considered since they influence the price of material containment and operation in commercial systems.

Sensible Heat Storage

Sensible heat can be hot or cold potential energy with respect to the ambient temperature, stored in a solid, liquid, or dual medium that consists of some combination of liquid and solid media. The internal energy change in sensible heat is dependent upon mass, specific heat, and temperature change:

$$\Delta U = mc_p(T_1 - T_2) \tag{7.1}$$

ΔU represents the change in internal energy of the material in kilojoules, m represents the mass of the material in kilograms, c_p is the specific heat capacity (kJ/kg·K) and T_1 and T_2 are the initial and final temperatures of the material, respectively, in Kelvin.

Sensible Heat Storage Materials

The material chosen for TES with sensible heat must be thermally stable and undergo no phase change during temperature extremes. The substance should also have a high heat capacity, high density, and an acceptably low vapor pressure. To be economically viable, it must be inexpensive. Several common sensible heat storage materials and some of their thermal properties are listed in Table 7.1.

Latent Heat

Latent heat is the amount of energy in the form of heat released or absorbed by a material during a change of state or phase transition such as solid to liquid (melting) or liquid to gas (vaporization). Latent heat energy storage is

TABLE 7.1

Physical Properties of Sensible Storage Materials

Storage Medium	Temperature (°C)		Average Density (kg/m³)	Average Heat Conductivity (W/mK)	Average Heat Capacity (kJ/kgK)	Volume-Specific Heat Capacity (kWh₁/m³)	Media Cost ($/kg)	Media Cost ($/kWh₁)
	Cold	Hot						
Solid media								
Sand–rock–mineral oil	200	300	1,700	1.0	1.30	60	0.15	4.2
Reinforced concrete	200	400	2,200	1.5	0.85	100	0.05	1.0
Solid NaCl	200	500	2,160	7.0	0.85	150	0.15	1.5
Cast iron	200	400	7,200	37.0	0.56	160	1.00	32.0
Cast steel	200	700	7,800	40.0	0.60	450	5.00	60.0
Silica fire bricks	200	700	1,820	1.5	1.00	150	1.00	7.0
Magnesia fire bricks	200	1,200	3,000	5.0	1.15	600	2.00	6.0
Liquid media								
Mineral oil	200	300	770	0.12	2.6	55	0.30	4.2
Synthetic oil	250	350	900	0.11	2.3	57	3.00	43.0
Silicone oil	300	400	900	0.10	2.1	52	5.00	80.0
Nitrite salts	250	450	1,825	0.57	1.5	152	1.00	12.0
Nitrate Salts	265	565	1,870	0.52	1.6	250	0.70	5.2
Carbonate salts	450	850	2,100	2.0	1.8	430	2.40	11.0
Liquid sodium	270	530	850	71.0	1.3	80	2.00	21.0

Source: Hermann, U., Geyer, M., and Kearney, D. 2002. *Overview of Thermal Storage Systems.* With permission.

attractive for materials that undergo a significantly high change in internal energy during a phase change.

Latent Heat Storage via Phase Change Materials

The phase change energy (heat of fusion or vaporization) of a material determines its thermal storage capacity as a phase change material (PCM). Molecular bonds of a PCM are broken when sufficient heat is applied. Their bonding energy gives PCMs their exceptional heat capacity. To be appropriate for latent heat storage, materials must exhibit a high heat of transition, high density, appropriate transition temperature, low toxicity, and long-term performance at low cost. For example, paraffin waxes and salt hydrates have

TABLE 7.2

Physical Properties of Latent Storage Materials

Phase Change Storage Medium	Temperature (°C)	Average Density (kg/m³)	Average Heat Conduc- tivity (W/mK)	Average Heat Capacity (kJ/kgK)	Volume- Specific Heat Capacity (kWh$_t$/m³)	Media Cost ($/kg)	Media Cost ($/kWh$_t$)
NaNo$_3$	308	2,257	0.5	200	125	0.20	3.6
KNO$_3$	333	2,110	0.5	267	156	0.30	4.1
KOH	380	2,044	0.5	150	85	1.00	24.0
Salt–ceramic (NaCo$_3$– BaCO3–MgO)	500–850	2,600	5.0	420	300	2.00	17.0
NaCl	802	2,160	5.0	520	280	0.15	1.2
Na$_2$CO$_3$	854	2,533	2.0	276	194	0.20	2.6
K$_2$CO$_3$	897	2,290	2.0	236	150	0.60	9.1

Source: Hermann, U., Geyer, M., and Kearney, D. 2002. *Overview of Thermal Storage Systems.* With permission.

high volumetric energy densities for small temperature swings, making them good latent heat storage materials.

The primary advantage of phase change materials is their ability to store energy at reduced temperatures and in reduced quantities. Furthermore, their high heat of fusion and other thermal properties allow freezing with little supercooling, no segregation, chemical stability, and a sharp melting point. One advantage is that some PCMs have low thermal conductivity in the solid state, making high heat transfer rates necessary during the freezing cycle. They are often flammable, requiring the addition of safety features to the design of storage vessels. In addition, latent heat is much more difficult to transfer than sensible heat. Table 7.2 shows the material properties of several common media used for latent heat storage. A TES designer should choose materials based on these specific characteristics.

Thermochemical Energy

Thermochemical energy is stored as the bond energy of a chemical compound. During a thermochemical reaction, atomic bonds are broken through a reversible chemical reaction and are catalyzed by an increase in temperature—which allows the energy to be stored. After thermochemical separation occurs, the constituents are stored apart until the combination reaction is desired. Recombination of the bonds between atoms releases the stored thermochemical energy.

TABLE 7.3

Thermochemical Storage Reactions

Reaction	$\Delta H°$ (kJ)	T′ (K)
$NH_4F(s) \leftrightarrow NH_3(g) + HF(g)$	149.3	499
$Mg(OH)_2(s) \leftrightarrow MgO(s) + H_2O(g)$	81.1	531
$MgCO_3(s) \leftrightarrow MgO(s) + CO_2(g)$	100.6	670
$NH_4HSO_4(l) \leftrightarrow NH_3(g) + H_2O(g) + SO_3(g)$	337.0	740
$Ca(OH)_2(s) \leftrightarrow CaO(s) + H_2O(g)$	109.3	752
$BaO_2(s) \leftrightarrow BaO(s) + ½O_2(g)$	80.8	1000
$LiOH(l) \leftrightarrow ½ Li_2O(s) + ½H_2O(g)$	56.7	1000
$CaCO_3(s) \leftrightarrow CaO(s) \, CO_2(g)$	178.1	1110
$MgSO_4 \leftrightarrow MgO(s) + SO_3(g)$	287.6	1470

Source: Wyman, C. March 1979. Thermal Energy Storage for Solar Applications: An Overview. SERI/TR-34-089, Solar Energy Research Institute, Golden, CO. With permission.

The primary advantages of thermochemical storage include high energy density and long term, low temperature storage capability. However, the thermochemical process is complex; the thermochemical materials are often expensive and can be hazardous [3].

Thermochemical Energy Storage

Since high density is important for energy storage, only reversible reactions with reactants and products that can be stored easily as liquids and solids are of practical interest. Reactions that produce two distinct phases such as a solid and a gas are desirable since the separation of products to prevent back-reaction is facilitated. Table 7.3 shows several common thermochemical storage reactions and their standard enthalpy change ($\Delta H°$) in kilojoules and turning temperature (T′) in Kelvin.

The turning temperature or T′ in Table 7.3 is defined as the temperature for which the equilibrium constant is one and is calculated using the ratio of the standard enthalpy change and the standard entropy change for the reaction. At this temperature, the reactants and products are present in approximately equal quantities. When $T > T′$, the endothermic storage reaction is favored, meaning heat is necessary for and absorbed during the reaction. Conversely, for $T < T′$, the exothermic reaction dominates—heat is a product of the reaction.

Choosing Storage Method

Many factors contribute to choosing an appropriate storage method and material for a TES application. Table 7.4 lists several options for solar thermal power and their appropriate storage materials.

TABLE 7.4

Options for TES in Solar Power Production

Options	Temp. (°C)	Storage Medium	Type
Small power plants and water pumps			
Organic Rankine	100	Water in thermocline tank or two tanks	Sensible
	300	Petroleum oil in thermocline tank	Sensible
Steam Rankine with organic fluid receiver	375	Synthetic oil with trickle charge	Sensible
Dish mounted engine generators (buffer storage only)			
Organic Rankine	400	Bulk PCM with indirect HX	Latent
Stirling and air Brayton	800	Bulk PCM with indirect HX	Latent
Advanced air Brayton	1370	Graphite	Sensible
		Encapsulated PCM	Latent
Large power plants (typically 3 to 8 hours of storage)			
Steam Rankine with organic fluid receiver	300	Petroleum oil in thermocline tank or two tanks, evaporation only	Sensible
		Petroleum oil and rocks (dual media in thermocline tank)	Sensible
Steam Rankine with water–steam receiver	300	Petroleum oil in thermocline tank or two tanks, evaporation only	Sensible
		Petroleum oil and rocks (dual media in thermocline tank)	Sensible
		Encapsulated PCM with evaporative HX	Latent
		Bulk PCM with indirect HX	Latent
		Bulk PCM with direct HX	Latent
		Pressurized water under or above ground	Latent
	540	Molten draw salt in thermocline tank or two tanks, superheat	Sensible
		Air and rocks	Sensible
		Bulk PCM with direct HX, evaporation stage	Latent

(Continued)

TABLE 7.4 (CONTINUED)

Options for TES in Solar Power Production

Options	Temp. (°C)	Storage Medium	Type
		Solid or liquid decomposition, evaporation stage	TC
Steam Rankine with molten draw salt receiver	540	Molten draw salt in thermocline tank or two tanks	Sensible
Steam Rankine with liquid metal receiver	540	Liquid sodium in one tank, mixed, buffer only	Sensible
		Liquid sodium in two tanks	Sensible
		Air and rocks	Sensible
Brayton with gas-cooled receiver	800	Refractory or cast iron in pressure vessel	Sensible
		Bulk PCM with indirect HX	Latent
		Solid or liquid decomposition	TC
Brayton with liquid-cooled receiver	800	VHT molten salt in two tanks	Sensible
		VHT molten salt and refractory (dual media) in thermocline tank	Sensible
	1100	Bulk glassy slag, liquid and solid bead storage, direct HX	Sensible, latent

Source: de Winter, F. 1990. *Solar Collectors, Energy Storage, and Materials*, MIT Press, Cambridge MA. With permission.

PCM = phase change material. HX = heat exchanger. VHT = very high temperature. TC = thermochemical.

Storage Systems

Two-Tank Direct Storage

In a two-tank direct storage system, the material used to store thermal energy is the same as the heat transfer fluid used to collect the thermal energy. The fluid is stored in two tanks: one at a high temperature and the other at a lower temperature. Fluid from the lower temperature tank flows through a solar collector or receiver, where solar energy heats it to the high temperature and it then flows back to the high temperature tank for storage. Fluid from the high temperature tank flows through a heat exchanger, where it generates steam for electricity production. The fluid exits the heat exchanger at a low temperature and returns to the low temperature tank.

Two-tank direct storage was used in early parabolic trough power plants at the Solar Electric Generating Station I and at the Solar Two power tower in California, as discussed later in this chapter. The two trough plants used mineral oil as the heat transfer and storage fluid; the Solar Two power tower used molten salt.

Molten Salt as Heat Transfer Fluid

Using molten salt at a solar field and in a TES system eliminates the need for expensive heat exchangers. This concept allows a solar field to operate at higher temperatures than systems using other common heat transfer fluids such as oils. Due to the elimination of heat exchangers and the reduction of heat transfer fluids, the use of molten salt as a heat transfer fluid substantially reduces the total cost of a TES system.

Unfortunately, molten salts freeze at relatively high temperatures—about 120 to 220°C (250 to 430°F). This means that special care must be taken to ensure that the salt does not freeze in the solar field piping. The Italian research laboratory, ENEA, and Sandia National Laboratories in the United States are currently developing new salt mixtures with the potential for freeze points below 100°C (212°F) to make molten salt much more manageable as a heat transfer fluid.

Two-Tank Indirect Storage

The two-tank indirect system functions in the same way as the two-tank direct system, but different fluids are used for heat transfer and storage. This system is used in plants where the heat transfer fluid is too expensive or not suited for use as the storage fluid. The storage fluid from a low temperature tank flows through an extra heat exchanger, where it is heated by the high temperature heat transfer fluid. The high temperature storage fluid then flows back to the high temperature storage tank. The fluid exits the heat exchanger at a low temperature and returns to the solar collector or receiver, where it is heated back to the high temperature. Storage fluid from the high-temperature tank is used to generate steam in the same manner as the two-tank direct system. The indirect system requires an extra heat exchanger and this adds cost to the system and decreases the overall TES efficiency. This system will be used in several parabolic power plants in Spain and has also been proposed for several U.S. plants that will use organic oil as the heat transfer fluid and molten salt as the storage fluid. Later in this chapter, Figure 7.11 illustrates Abengoa Solar's Solana plant in which a two-tank indirect storage system is used with oil as the heat transfer fluid and molten salt as the storage material.

Single-Tank Thermocline Storage

A single-tank thermocline storage system stores thermal energy in a solid medium—usually silica sand—located in a single tank. At any time during operation, a portion of the medium is at high temperature and a portion is at low temperature. The hot and cold temperature regions are separated by a temperature gradient or thermocline. High temperature heat transfer fluid flows into the top of the thermocline and exits the bottom at low temperature. This process moves the thermocline downward and adds thermal energy to the system for storage. Reversing the flow moves the thermocline upward and removes thermal energy to generate steam and electricity. Buoyancy effects create thermal stratification of the fluid in the tank, which helps stabilize and maintain the thermocline. Figure 7.3 shows the basic thermocline storage tank concept in which hot and cold materials are stored inside the tank.

The thermocline technology has proven advantageous because the reduction of materials used for constructing the tank and storing heat decreases cost and energy input. Using a solid storage medium for only one tank reduces the cost relative to the two-tank systems. The thermocline system was demonstrated at the Solar One power tower, where steam was used as the heat transfer fluid and mineral oil was used as the storage fluid. However, this technology is still undergoing development and requires more research before it is economically and technically viable.

Storage Vessel Design

Tank Geometry

The cylinder is the most practical and common geometry for a storage tank. Spherical storage tanks are also common for limited applications. For instance, spherical vessels are typically used underground or supported by columns for gravity pressured supply; above-ground tanks are most often cylindrical because of construction practicalities. These geometries will be compared in this section. Figure 7.4 indicates the parameters used to characterize them.

For a given volume, spherical geometry presents the least surface area—a desirable factor for minimizing the materials and area for heat transfer to the surrounding environment. The surface area-to-volume ratio for a sphere is a constant equal to $3/r$, so the size of a spherical vessel is determined by the volume of storage material needed. The cylindrical shape with the least surface area is one whose height, h, is equivalent to its diameter; the surface area-to-volume ratio is $3/2r$. In the best case where h is twice the radius, a cylinder has 1.5 times the surface area of the sphere and just $2/3$ of its volume. Table 7.5 shows how these two geometries compare.

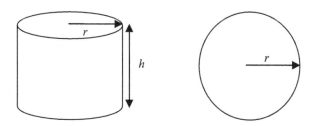

FIGURE 7.4
Cylindrical and spherical tank geometries.

TABLE 7.5

Comparison of Surface Area and Volume of Cylindrical Tank
Geometries Based on Radius

TES Geometry	Surface Area	Volume	Surface Area/Volume
Sphere	$4\pi r^2$	$\dfrac{4\pi r^3}{3}$	$\dfrac{3}{r}$
Cylinder of variable height	$2\pi r(r+h)$	$\dfrac{rh}{2(r+h)}$	$\pi r^2 h$
Optimized cylinder (where $h=2r$)	$6\pi r^2$	$2\pi r^3$	$\dfrac{3}{2r}$

Figure 7.5 illustrates the relations in Table 7.5: the surface areas and volumes of cylindrical and spherical tanks depend on the radius of each vessel and as the radius increases, the areas and volumes of both geometries increase. At large radii, the cylindrical geometry results in a much greater surface area and volume.

These simple geometric comparisons lead one to opt for a cylindrical geometry for a TES tank. Additionally, the materials used in a storage tank are much more difficult to customize for a spherical tank and a spherical tank is far more difficult to construct. Ensuring mechanical support for a spherical tank is also more difficult. For these reasons, spherical tanks are often more practical for underground use. The requirement for more customized materials will inevitably increase the cost of an entire system and the time needed to complete it. Figure 7.6 shows the longitudinal cross-section view of a representative cylindrical storage tank with heat exchangers.

Tank

Materials

The materials used for a tank are critical to its performance. The tank materials will exert almost as much influence on the efficiency of heat storage as the

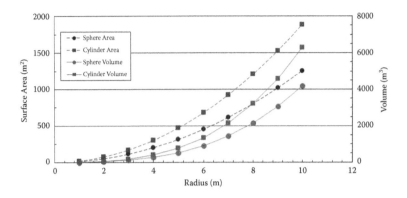

FIGURE 7.5
Comparison of surface area and volume for both cylindrical and spherical storage tank geometries.

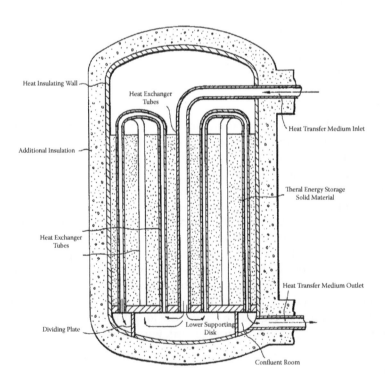

FIGURE 7.6
Longitudinal cross-sectional view of cylindrical storage tank. (From Anzai, S. et al. March 30, 1977. Thermal Energy Storage Tank. Patent 4,088,883 assigned to Japan Agency of Industrial Science & Technology, Tokyo.)

storage material chosen. Many tank materials have been tested and undergone industrial experience. The consensus in the industry is to use concrete and steel with various insulation materials.

Steel is optimal for very large storage tanks in which large volumes of fluid exert high pressure on the storage container; appropriate design is critical. Steel tanks have high strength and can be welded on site. The interior and exterior of a steel tank must be coated to ensure corrosion resistance. Concrete is recommended for unpressured systems. It is a low cost material capable of storing large volumes contents. Care must be taken with the surface treatment to keep fluids from seeping into the concrete. Fiberglass is often used because of its high corrosion resistance. However, fiberglass is expensive and difficult to connect to because of its formed design. Plastic is used for thermal energy storage applications with low temperatures and smaller volumes and can be an economical choice [18].

Material choice should be based on tank area. More specifically, a designer must consider leak potential, conduction into soil, and accessibility to the bottom of the tank. For the inside of the tank, additional leak protection, heat loading, drains, and controls for high and low temperature conditions are important factors. The tank exterior must be resistant to ultraviolet radiation and temperature effects and should have insulating and waterproofing properties. Weight bearing strength of all areas of a tank should be considered because the outermost bearings carry the load, fluid, insulation, saddles, and fittings. These design criteria are detailed further in the next section covering stress.

Stress and Strain

Mechanical Stress

Plane stress must be accounted for in the design of a tank. The stress tangent to the plane of a tank's walls is called hoop stress, commonly represented as σ, and is a function of the pressure or p applied by the fluid, the radius or r of the tank, and the thickness or t of the wall material. In this case, p is the gauge pressure of the fluid as compared to the pressure outside of the tanks. Cylindrical hoop stress is characterized as

$$\sigma = \frac{pr}{t} \tag{7.2}$$

Cylindrical longitudinal stress is characterized:

$$\sigma = \frac{pr}{2t} \tag{7.3}$$

For a sphere, the plane stress is the same as cylindrical longitudinal stress. The equations above for plane stress in a tank show that the radius of the tank and the thickness of the tank material are the effective parameters for safe design. Another important parameter is tank temperature.

Thermal Strain

Since the materials used in thermal energy storage undergo very high temperatures, one must fully understand the effect of temperature on the materials used. The thermal expansion of a material must be known in order to calculate how much tension the tank will undergo when it expands due to elevated storage temperatures. Thermal strain is calculated as

$$\Delta x = \alpha \Delta TL \tag{7.4}$$

The combination of mechanical stress and thermal strain must be considered when designing a storage tank to ensure the resulting stress is well below the ultimate strength of the material [3].

Heat Loss from Storage Vessels and Insulation

Heat loss to the environment is a function of a storage vessel's surface area-to-volume ratio [10]. Insulation is an effective way to reduce heat loss to the environment. The appropriate design of insulation in a storage vessel depends on a balance of the ambient heat loss and cost of insulation. Since heat loss to the environment depends on the surface area-to-volume ratio, it will be considered separately for cylindrical and spherical vessels.

Heat Loss from Cylindrical Vessels

The heat loss through insulation of a cylindrical tank [5] can be calculated:

$$Q = \frac{k2\pi L(T_s - T_a)}{\ln(R_2 / R_1) + (k / h_t R_2)} \tag{7.5}$$

Equation 7.5 allows calculation of heat loss for a tank based on the conditions in Table 7.6 which gives the variable meanings, units and representative values of each variable.

To quantify whether a cylindrical tank should be insulated, Table 7.7 and Figure 7.7 show the insulation costs and insulation thickness as a function of heat loss, assuming a 42% efficient Rankine cycle and an electricity cost of $0.10/kWh. These data show the economic effectiveness of insulating a storage tank. Applying just 10 cm of insulation dramatically reduces ambient heat loss and returns the investment in 3 years.

TABLE 7.6

Variables and Properties Contributing to Heat Loss
through Cylindrical Insulated Tank

Symbol	Variable	Unit	Value
Q	Heat loss	kW	0.581
K	Coefficient of Conduction	W/mK	0.035
L	Cylinder length	m	5
T_s	Surface temperature	C	100
T_a	Ambient temperature	C	0
R_1	Tank radius	m	2.5
R_2	Insulation radius	m	3
h_t	Radius and conversion coefficient	W/m²K	15

Source: Brumleve, T.D. 1974. Sensible Heat Storage in
Liquids, Sandia Laboratories Energy Report, SLL-
73-0263. Livermore, CA

TABLE 7.7

Monetary Value of Insulating Cylindrical Tank

Heat Loss (kW)	Insulation Thickness (m)	R2	Insulation Volume (m³)	Insulation Cost ($)	Cost of Heat Loss ($/Year)	Payback Time (Years)
18.85	0	2.5	0.00	0	6,935	Infinite
2.428	0.1	2.6	8.01	2,648	893	3
1.321	0.2	2.7	16.34	5,400	486	11
0.917	0.3	2.8	24.98	8,256	337	24
0.708	0.4	2.9	33.93	11,215	260	43
0.581	0.5	3	43.20	14,279	214	67

Source: Brumleve, T.D. 1974. Sensible Heat Storage in Liquids, Sandia Laboratories
Energy Report, SLL-73-0263. Livermore, CA

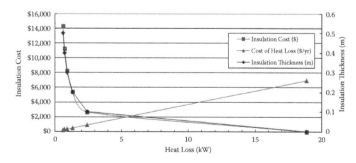

FIGURE 7.7
A plot of cylindrical tank heat loss versus insulation thickness shows the value of insulation.

Heat Loss from Spherical Vessels

T. Brumleve developed relations to describe sensible heat loss from spherical thermal energy storage volumes. The heat loss from a spherical tank, Q_L, with constant temperature, T_{avg}, is shown in Equation (7.6) [5]. The heat stored in the tank, Q_S, is given by Equation (7.7) [4].

$$Q_L = \frac{4k\pi R^2 t}{L}(T_{avg} - T_C) \tag{7.6}$$

$$T_{avg} = \frac{T_H + T_C}{2} \tag{7.7}$$

Typical values of each parameter for a 20 m diameter spherical energy storage tank filled with water are shown in Table 7.8. These values can be used with Equations (7.6) and (7.7) to find the heat loss from a tank and the amount of heat stored in the tank over time. Figure 7.8 shows heat loss from a spherical tank as a function of time and the tank's thermal capacity. Solar Two (discussed later in this chapter) is a 10 MW$_e$ power tower that stores molten salt in hot and cold tanks. The hot tank is 11.6 m in diameter and 8.4 m tall. It was designed to store 105 MWh$_t$ of thermal energy and supply 3 hours of full scale electricity production. Solar Two's heat loss to the environment was measured from the hot and cold tanks, steam generator, and receiver sumps. A total of 185 kW of heat loss from these components is equivalent to 2% of the collected thermal energy on a winter day. Table 7.9 is a demonstration of where calculated and measured heat losses arise within a thermal energy storage system.

TABLE 7.8

Thermal Storage Parameters for Spherical Tank

Parameter	Sample Value	Unit	Description
Q_L		J	Heat loss to environment
Q_S		J	Heat stored in tank
K	0.035	W/m×K	Thermal conductivity of insulation
R	10	M	Radius of tank
T		S	Storage time
L	0.5	M	Thickness of insulation
T_{ave}	368	K	$(T_H+T_L)/2$ for diurnal storage
T_{amb}	293	K	Temperature of ambient medium
T_H	393	K	Temperature of hot fluid
T_C	343	K	Temperature of cool fluid
C_p	4217	J/kg×K	Specific heat of water
ρ	958	kg/m³	Density of water

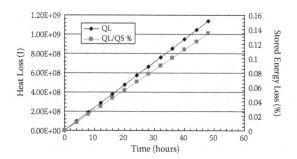

FIGURE 7.8

Heat loss from spherical tank compared to the energy stored in the tank.

TABLE 7.9

Thermal Losses for Major Components of Solar Two TES

Major Equipment	Calculated Thermal Loss (kW)	Measured Thermal Loss (kW)
Hot tank	98	102
Cold tank	45	44
Steam generator sump	14	29
Receiver sump	13	9.5

Source: National Renewable Energy Laboratory. 2000. *Survey of Thermal Storage for Parabolic Trough Power Plants.* NREL/SR-550-27925. Washington, D.C.

Economics of Thermal Energy Storage Systems

Thermal storage is now the least expensive clean energy storage option available. However, the cost can only be reduced further through deployment and economies of scale. TES is economical when one or more of the following conditions exists:

- A utility experiences high demand.
- A utility provider charges time-of-use rates (some charge more for energy use during peak periods of day and less off-peak).
- Daily loads vary greatly.
- Loads are of short duration or are infrequent or cyclical.
- Cooling equipment has trouble handling peak loads.
- Rebates are available for load shifting to avoid peak demand.

TES systems are installed for two major reasons: lower initial project costs and lower operating costs. Initial cost may be lower because distribution temperatures are lower and equipment and pipe sizes can be reduced. Operating costs may be saved due to smaller compressors and pumps along with reduced time-of-day or peak demand utility costs. The economics of thermal storage is very specific to a particular site and system. A feasibility study is generally required to determine the optimum design for a specific application. Several examples of effective TES installations that cost less than conventional alternatives and also provided significant energy and energy cost reductions exist.

TES projects often profit from unexpected benefits that are secondary to the primary reason for an action. For example, a well designed TES air conditioning application may experience reduced chiller energy consumption, lower pump horsepower, smaller pipes, high reliability, better system balancing and control, and lower maintenance costs.

Peak Shaving

High peak summertime loads make energy costs exceedingly expensive to consumers and providers. The industry meets these peak loads with low-efficiency peaking power plants, usually gas turbines, that represent lower capital costs but higher fuel costs and greater environmental impacts than arise from other energy sources. A kilowatt hour of electricity consumed at night can be produced at much lower marginal cost than a kilowatt hour consumed during peak times. Thermal energy storage allows a solar plant to potentially "shave" peak loads.

Cost to Energy Provider

Energy storage allows a plant operator to maximize profits by manipulating energy prices, peak shaving, reducing intermittence, and increasing plant utilization. During periods of low hourly power prices, the operator can forego generation and dump heat into storage; at times of high prices, the plant can run at full capacity even without sun. Solar generating capacity with heat storage can make other capacities unnecessary. The ability of thermal solar plants to use heat energy storage to keep electric output constant: (1) reduces the costs associated with uncertainty surrounding power production and (2) relieves concerns about electrical interconnection fees, regulation service charges, and transmission tariffs. Solar plants equipped with heat storage have the ability to increase overall annual generation by "spreading" solar radiation to better match plant capacity.

Cost of Storage Implementation

The primary costs for a TES system are the storage material, heat exchangers, storage tank, and insulation. Washington State University conducted a case

TABLE 7.10

Monthly Demand Savings from Use of Thermal
Storage to Shift Chiller Operation to Night Time*

Utility Demand Rate ($)	Cost /Ton Hour of Cooling ($)	Monthly Demand Savings ($)
6/kW	4.20	1,008
12/kW	8.40	2,016

Source: Washington State University Cooperative
Extension. 2003. *Energy Efficiency Fact Sheet.*
WSUCEEP00-127.

*Assumes 300-ton chiller operating at average load—80% of
full load or 240 tons. Chiller efficiency assumed at 0.7
kW/ton. Chiller operation assumed at night when building is unoccupied so that peak demand is not increased
by chiller operation.*

TABLE 7.11

Comparison of Thermal Storage Media

Parameter	Chilled Water	Ice
Chiller cost	$200 to 300/ton	$200 to 1,500/ton
Storage tank cost	$30 to 150/ton hour	$20 to 70/ton hour
Storage volume	6 to 20 foot³/ton hour	2.5 to 3.3 foot³/ton hour
Chiller efficiency	5 to 6 COP	2.7 to 4 COP

Source: Washington State University Cooperative Extension. 2003.
Energy Efficiency Fact Sheet. WSUCEEP00-127.

study in which the Dallas Veterans Affairs Medical Center installed a 24,628
ton/hour chilled water TES system. The system resulted in a reduction in
demand of 2,934 kW and a reduction in annual electricity cost of $223,650
[8]. The local utility provided $500,000 of the $2.2 million required for design
and installation. Savings resulting from installation of the thermal storage
technology will allow the VA to recoup its investment within 7 years [8].
Tables 7.10 and 7.11 detail more of the study findings regarding savings for
the energy provider and the costs of specific TES components for solar chilling. One can assume that the costs presented for solar chilling storage are
comparable for solar heating storage.

Cost to Consumer

Figure 7.9 shows a comparison of the theoretical monthly load versus time
of day first without, then with energy storage. This illustrates the economic
advantage of energy storage coupled with solar power because at times of

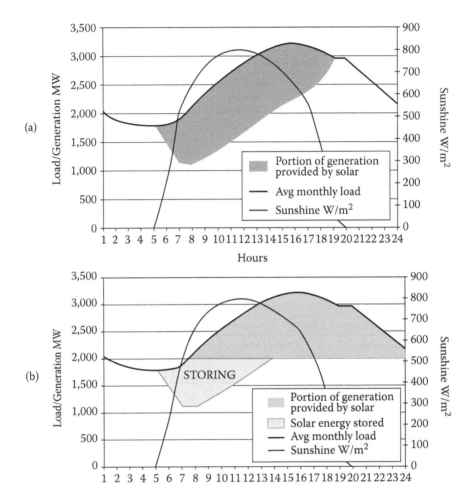

FIGURE 7.9
Theoretical load and generation versus time of day for utility power (a) without storage and (b) with storage.

peak power usage, a utility is able to better match the demand curve with stored energy and then offer cheaper electricity at peak times.

Figure 7.10 shows data generated by Arizona Public Service (APS) on a summer day. The available solar energy output with thermal energy storage matches the load almost exactly, whereas the solar output without thermal energy storage begins to decline just as the load peaks. Energy is most expensive to consumers when the load peaks, around 2 p.m., due to simple supply and demand principles. By adding storage to a system, a utility can afford to decrease peak energy costs to consumers.

Heat equivalents up to two days of power plant operation can be stored for high demand–high price electricity periods that can occur 3–4 hours after

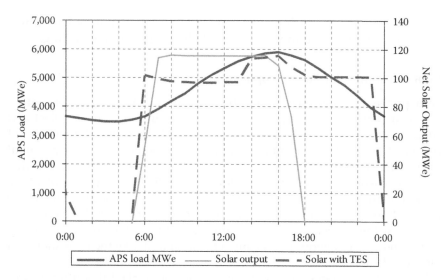

FIGURE 7.10

Arizona Public Service's load output profile from a parabolic trough compared with the solar output with and without TES on a summer day.

dark. Additionally, two days of energy storage could mean that weekend energy can be stored for use when industrial demand resumes on Mondays. The typical cost of concentrated solar power (CSP) is estimated by the U.S. National Renewable Energy Laboratory (NREL) in Golden, Colorado, at about $0.17/kWh. To compete with current fossil fuel prices, the cost of CSP must be reduced to about $0.05/kWh. This emphasizes the need for continued research and development of thermal solar power systems, particularly TES.

TES Applications

Practical applications of solar thermal energy storage include, but are not limited to

- Space heating and cooling
- Domestic water heating
- Industrial and agricultural process heating
- Solar cooking
- Small power plants and water pumps
- Dish engine generators (buffer storage)
- Large concentrating solar power plants (typically 3 to 12 hours of storage)

This section will focus on the applications of TES in concentrating solar power plants and building and industrial process heat and seasonal heating.

Concentrating Solar Power Applications

Perhaps the most pertinent application of TES is in large scale concentrating solar power (CSP) plants. CSP technologies use mirrors to reflect and concentrate sunlight onto receivers in a solar field that collect the energy and convert it to heat. One advantage of parabolic trough CSPs is their potential for storing solar thermal energy to use during nonsolar periods and dispatch it when most needed. TES allows parabolic trough power plants to achieve higher annual capacity factors—from 25% without thermal storage up to 70% or more with it. The heat from the concentrated radiation is used to increase the temperature of a working heat transfer fluid to approximately 400°C and carries energy to a steam generation unit, like the ones shown in Figures 7.11 and 7.12.

Figure 7.11 illustrates a concentrating solar power plant that utilizes two-tank indirect TES. The expansion of steam through the turbine produces mechanical torque for electric generation [6]. The steam cycle next undergoes condensation and heat rejection before recycling to gain heat from exchange with the absorber heat transfer fluid [6]. This vapor power production is described by a four-step Rankine cycle and is shown in Figure 7.12 [6]. The Rankine cycle involves (1) isentropic compression of water by a pump, (2) constant pressure heat addition in the steam generator, (3) isentropic expansion

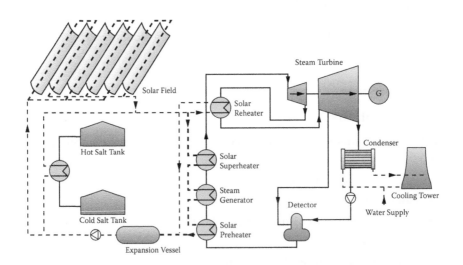

FIGURE 7.11
Diagram of Abengoa Solar's Solana power plant in Gila Bend, Arizona. (From Abengoa Solar: Our Projects. Solana. http://www.abengoasolar.com/corp/web/en/our_projects/solana/index.html. With permission.)

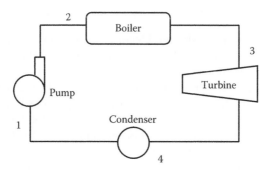

FIGURE 7.12
Rankine steam cycle power generation. (From Cengel, Y.A., and Boles, M.A. 1994. *Thermodynamics: An Engineering Approach.* McGraw Hill, New York. With permission.)

TABLE 7.12

Primary CSP Technologies

Technology	Parabolic Trough	Power Tower	Stirling Dish
Description	Linear trough solar concentration to steam turbine power cycle	Heliostat mirror array with central tower receiver	Point focus, fluid expansion, electricity generation in Stirling engine
Sun to electric efficiency	14%	16%	20%
Concentration factor	100 suns	1000 suns	3000 suns
Heat transfer fluid temperature	400°C oil	600°C molten salt	725°C oil
Space requirement	5 acres/MW	8 acres/MW	4 acres/MW
Advantage	Most developed CSP (SEGS)	TES easily incorporated	Modular, low investment cost
Disadvantage	Cooling water required for turbine cycle cooling	Large area requirement	Difficult to incorporate TES

of steam through the turbine, (4) constant pressure condensation of the steam back into water.

The three proven CSP technologies for collection, concentration, and generation of solar power are the parabolic trough, the solar power tower, and the Stirling dish engine. Each has a different geometry and associated concentration of solar energy on the absorber. Parabolic troughs and power towers are the most advanced systems for large scale power production and incorporation of TES. Table 7.12 compares properties of these CSP concepts.

Existing Large-Scale Solar Thermal Energy Storage Systems

Solar One and Solar Two Mojave Desert, California — The first commercial concentrating solar power tower system was a 10 MW pilot plant called Solar One that included a TES tank used to test the performance of thermal storage within oil, rocks, and sand media. Eventually, molten salt at 565°C was determined to be the best storage material. The project produced 10 MW of electricity using 1,818 mirrors of 40 m² each, with a total area of 72,650 m². Solar One was completed in 1981 and operational from 1982 to 1986.

In 1995, Solar One was converted into Solar Two by adding a second ring of 108 larger (95 m²) heliostats around the existing Solar One for a total of 1,926 heliostats with an area of 82,750 m². The new Solar Two plant could also produce 10 MW. Solar Two used molten salt (a combination of 60% sodium nitrate and 40% potassium nitrate) as an energy storage medium. It was decommissioned in 1999 and converted by the University of California Davis into an Air Cherenkov Telescope in 2001 to measure gamma rays entering the atmosphere.

Solar Energy Generating Systems (SEGS), Mojave Desert, California—In 1985, Luz International built the first SEGS at Kramer Junction in the Mojave Desert. There are now nine SEGS plants and together they comprise the world's largest solar energy generation facility. SEGS I is a 13.4 MW plant with 3 hours of thermal storage capacity using Caloria mineral oil.

Table 7.13 compares the SEGS I and Solar One and Two systems along with other industrial TES applications. Further development of storage technologies such as TES is required for baseload power to be provided by renewable energy sources.

Building and Industrial Process Heat

Thermal energy storage techniques can be used on a smaller scale in buildings and industrial processes. Tables 7.14 and 7.15 list common applications for TES within building and process heating applications, respectively. For example, cleaning processes are required in many industries. Cleaning bottles, cans, kegs, and process equipment consumes most of the energy used by food industries. Metal treatment (galvanizing, anodizing, painting, etc.) plants use energy to clean parts and surfaces. Textile plants and laundries clean fabrics, and service stations clean cars. All these operations require warm (about 60 to 100°) water and provide excellent applications for TES. The storage and integration into existing heat supply systems is rather easy because they already have storage tanks for water that serve as the main medium [16].

Seasonal Heating

Seasonal heating is an application of TES designed to retain heat deposited during the hot summer months for use in colder winter weather. The heat is

TABLE 7.13

Industrial Applications of Thermal Energy Storage

Project	Type	Storage Medium	Cooling Loop	Nominal Temperature (°C)		Storage Concept	Tank Volume (m³)	Thermal Capacity (MWh$_t$)
				Cold	Hot			
Irrigation pump, Coolidge, AZ, USA	Parabolic trough	Oil	Oil	200	228	One tank, thermocline	114	3
IEA-SSPS Almeria, Spain	Parabolic trough	Oil	Oil	225	295	One tank, thermocline	200	5
SEGS I, Daggett, CA, USA	Parabolic trough	Oil	Oil	240	307	Cold tank		
Hot tank	4160 4540	120						
IEA-SSPS Almeria, Spain	Parabolic trough	Oil, cast iron	Oil	225	295	Dual, medium tank	100	4
Solar One, Barstow, CA, USA	Central receiver	Oil, sand, rock	Steam	224	304	Dual, medium tank	3460	182
CESA-1 Almeria, Spain	Central receiver	Liquid salt	Steam	220	340	Cold tank		
Hot tank	200 200	12						
THEMIS, Targasonne, France	Central receiver	Liquid salt	Liquid salt	250	450	Cold tank		
Hot tank	310 310	40						
Solar Two, Barstow, CA, USA	Central receiver	Liquid salt	Liquid salt	275	565	Cold tank		
Hot tank	875 875	110						

Source: National Renewable Energy Laboratory. 2000. *Survey of Thermal Storage for Parabolic Trough Power Plants.* NREL/SR-550-27925. Washington, D.C.

typically captured using solar collectors, although other energy sources may be used separately or in parallel. Seasonal thermal storage can be divided into three broad system types: low temperature, warm temperature multiseasonal, and high temperature systems. Low temperature systems use the soil adjoining a building as a seasonal heat store, drawing on the stored heat for space heating. This heat store often reaches temperatures similar to the average annual air temperature [14]. Such systems can also be seen as extension of buildings although this design involves some simple but significant differences not found in traditional buildings. Warm temperature interseasonal

TABLE 7.14

TES Building Applications

Application	Temperature (°C)	Storage Medium	Type
Space Heating: Active Solar Residential Scale, Diurnal Storage			
Air heating collectors	60	Rock beds	Sensible
	30 to 40	Encapsulated PCM	Latent
		Bulk solid / solid PCM	Latent
Liquid heating collectors	50	Stratified water in thermocline tank	Sensible
		Moistened replaced earth in lined pit, indirect HX	Sensible
		Bulk PCM with indirect HX	Latent
	30 to 40	Bulk PCM with direct HX and insoluble fluid	Latent
		Encapsulated PCM	Latent
Large-scale collectors (yearly averaging storage)	50 to 90	Water in tanks, excavated pits, mined caverns	Sensible
		Natural and artificial aquifers	Sensible
		Undisturbed earth, clay, rock	Sensible
Passive solar	30 to 45	Water walls	Sensible
		Trombe masonry walls	Sensible
		Encapsulated PCM in irradiated tubes or panels	Latent
		Dispersed PCM in architectural components	Latent
		Dispersed solid–solid PCM in architectural components	Latent
Space Cooling: Closed Cycle Systems			
Hot side storage for Rankine and liquid absorption chillers	90 to 120	Water in vented or pressurized tanks	Sensible
		Bulk PCM with indirect HX	Latent

TABLE 7.14 (CONTINUED)

TES Building Applications

Application	Temperature (°C)	Storage Medium	Type
		Bulk form-stable cross-linked polymer PCM, direct HX	Latent
Cold side storage for Rankine, liquid absorption, and zeolite chillers	5 to 10	Water in naturally stratified tank	Sensible
		Water in multiple, partitioned, or flexibly divided tanks	Sensible
		Encapsulated PCM	Latent
		Bulk PCM with indirect HX	Latent
		Bulk PCM with direct HX and immiscible fluid	Latent
Open cycle systems (dessicant chillers)	90	Rock beds for solid adsorption systems	Sensible
	60	Strong salt solution	TC
Space heating and cooling (thermochemical heat pumps with reactant storage)	100 to 130	Intermittent solid–vapor systems	Latent
		Continuous or intermittent liquid–vapor systems	Latent
Domestic Hot Water			
Naturally stratified deliverable water in thermocline tank	55	Supplementary PCM encapsulated bulk between double walls of water tank	Latent

Source: de Winter, F. 1990. *Solar Collectors, Energy Storage, and Materials*, MIT Press, Cambridge, MA. With permission.

heat stores also use soil to store heat, but employ active mechanisms of solar collection in summer to heat thermal banks in advance of the heating season [14]. High temperature seasonal heat stores are essentially extensions of a building's heating, ventilation, air conditioning, and water heating systems. Water is normally the storage medium, stored in tanks at temperatures that

TABLE 7.15

TES Process Heating Applications

Options for Industrial and Agricultural Process Heating	Temperature (C)	Storage Medium	Type
Process hot water	60 to 90	Deliverable water in thermocline tank or two tanks	Sensible
Low pressure steam	100 to 130	Pressurized water with indirect HX	Sensible
		Petroleum oil in thermocline tank or two tanks	Sensible
		Pressurized water delivered as steam	Latent
		Continuous liquid –vapor TC heat pump	TC
100-pound steam	170	Petroleum oil in thermocline tank or two tanks	Sensible
		Pressurized water delivered as steam	Latent
High pressure saturated steam	300	Petroleum oil in thermocline tank or two tanks	Sensible
		Petroleum oil/rocks (dual media) in thermocline tank	Sensible
		Encapsulated PCM with evaporative HX	Latent
		Bulk PCM with indirect HX	Latent
		Pressurized water under or above ground	Latent
Crop drying (air heating)	50 to 70	Rock beds	Sensible
		Liquid desiccant, yearly averaging	TC

TC = Thermochemical

Source: de Winter, F. 1990. *Solar Collectors, Energy Storage, and Materials*, MIT Press, Cambridge, MA. With permission.

can approach boiling [14]. Phase change materials and advanced soil heating systems are occasionally used instead. For systems installed in individual buildings, additional space is required to accommodate the storage tanks.

In all cases, effective above-ground insulation and heavy insulation of the building are required to minimize heat loss from the building, and hence the losses from heat that needs to be stored and used for space heating. Despite the differences in design, low temperature systems tend to offer simple and relatively inexpensive implementations that are less vulnerable to equipment failure [14]. They do, however, require a building site to be clear of the water table, bedrock, and existing buildings, and are limited to temperate or warmer climate zones and space heating operations. High temperature systems share the same vulnerabilities as conventional space and water heating systems due to their mechanical and electrical components but also allow greater control [13] and can be employed in colder climates.

References

1. Abengoa Solar: Our Projects. Solana. http://www.abengoasolar.com/corp/web/en/our_projects/solana/index.html
2. Arizona Public Service. Solana's Technology: Arizona's Largest Solar Power Plant. http://www.aps.com/main/green/Solana/Technology.html
3. Beckmann, G., and Gilli, P.V. 1984. *Thermal Energy Storage: Basics, Design, and Applications to Power Generation and Heat Supply.* Springer, Heidelberg.
4. Beer, F.P., and Johnston, E.R. 2001. *Mechanics of Materials*, 3rd ed. McGraw Hill, New York.
5. Brumleve, T.D. 1974. Sensible Heat Storage in Liquids. Energy Report 73-0263. Sandia National Laboratories, Albuquerque, NM.
6. Cengel, Y.A., and Boles, M.A. 1994. *Thermodynamics: An Engineering Approach.* McGraw Hill, New York.
7. de Winter, F. 1990. *Solar Collectors, Energy Storage, and Materials*, MIT Press, Cambridge, MA.
8. Washington State University Cooperative Extension. 2003. Energy Efficiency Fact Sheet. WSUCEEP00-127.
9. Hermann, U., Geyer, M., and Kearney, D. 2002. *Overview of Thermal Storage Systems.*
10. Krieth, F., and Bohn, M. 2001. *Principles of Heat Transfer*, 6th ed. Brooks Cole, New York.
11. National Renewable Energy Laboratory. 2000. *Survey of Thermal Storage for Parabolic Trough Power Plants.* NREL/SR-550-27925. Washington, D.C.
12. National Renewable Energy Laboratory. 2008. TroughNet: Parabolic Trough Thermal Energy Storage Technology. http://www.nrel.gov/csp/troughnet/thermal_energy_storage.html
13. Owens, B. 2003. The Value of Thermal Energy Storage. Platts Research & Consulting.
14. Paksoy, H.O. 2007. *Thermal Energy Storage for Sustainable Energy: Consumption Fundamentals, Case Studies, and Designs.* NATO Science Series, Springer, Heidelberg,
15. Anzai, S. et al. March 30, 1977. Thermal Energy Storage Tank. Patent 4,088,883 assigned to Japan Agency of Industrial Science & Technology, Tokyo.

16. International Energy Agency. Solar Heat for Industrial Processes. Task 33/Task IV. Solar Heating & Cooling Programme, SolarPACES. Newsletter No. 1.

17. Turchi, C. July 21, 2008. Thermal Energy Storage for Concentrating Solar Power Plants. National Renewable Energy Laboratory.

18. U.S. Department of Energy, *Active Solar Thermal Design Manual*, Contract EG-77-C-01-4042.

19. U.S. Department of Energy. Energy Efficiency and Renewable Energy. http://www1.eere.energy.gov/solar/images/parabolic_troughs.jpg

20. Wyman, C. March 1979. Thermal Energy Storage for Solar Applications: An Overview. SERI/TR-34-089, Solar Energy Research Institute, Golden, CO.

8

Natural Gas Storage

Kent F. Perry

CONTENTS

Introduction

The first recorded natural gas storage facility was a depleted gas reservoir converted in 1915 in Welland County, Ontario, Canada. Storage plays a key role in the U.S. natural gas industry, allowing it to smooth the differences between production patterns and gas demand and minimize the costs of transmission and distribution systems. Underground storage for natural gas is used for balancing supply and demand over a defined period to compensate for season-related (summer and winter) fluctuations to ensure that natural gas is available without delay for consumers (Beckman et al., 1995).

Without storage, production deliverability and pipeline capacity in the United States and Canada would need to be greatly increased to meet peak winter demand. As a result, storage provides significant incremental value to the natural gas industry. Natural gas may be stored in several different ways. It is most commonly held underground, under pressure, in (1) depleted reservoirs in oil and/or gas fields, (2) aquifers, and (3) salt cavern formations.

Each type has its own physical characteristics (porosity, permeability, retention capability) and economics (site preparation costs, deliverability rates, cycling capability) that govern its suitability to particular applications. Two of the most important characteristics of an underground storage reservoir are its capability to hold natural gas for future use and the rate at which gas can be withdrawn—its deliverability rate. The amount of gas in storage is another factor influencing short-term gas prices. Hence, storage capacity and usage are major concerns for the gas market.

Recent changes in the market for natural gas are altering the role and economics of natural gas storage. The development of market hubs has increased the importance of strategically located storage fields and expanded the types of services available by storage providers. Changes in pipeline rate structures and the increased importance of pipeline capacity discounting and capacity release have changed the economics of many storage facilities. In addition, changes in demand, particularly the growth in gas demand for electric generation, affect the overall load factors and significantly impact storage requirements (Figure 8.1) (Energy and Environmental Analysis, 2000).

Historical Development of Underground Natural Gas Storage

The first recorded natural gas storage facility was a converted depleted gas reservoir in Welland County, Ontario, Canada. The first storage field in the United States began operation in 1916 at Zoar Field near Buffalo, New York; it remains in service today. The success associated with providing additional supplies during peak demands by storing gas during summers in a depleted gas field made good sense to distributors and utilities in the Northeast, and other fields were developed gradually.

After World War II, due to increasing demands for energy to fuel post-war production and expansion, new, large diameter, long distance natural gas pipelines were laid to connect supply areas of Oklahoma, Kansas, Louisiana, and Texas to the large population centers of the Midwest and Northeast.

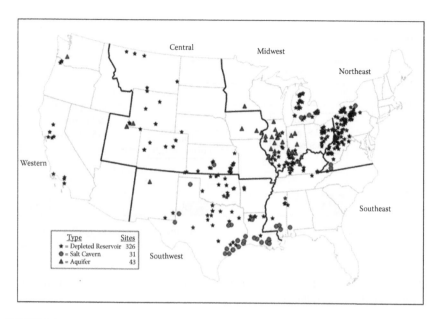

FIGURE 8.1
Locations and types of underground storage fields in the United States. (From U.S. Department of Energy.)

The industry recognized that new storage would be needed to serve these regions with weather-sensitive loads. Without new storage capabilities, the pipeline sizes would exceed the abilities of the steel industry of the 1950s to manufacture them. The alternative of laying numerous small lines was determined to be cost prohibitive.

From 1950 to 1965, the number of new gas storage fields increased dramatically. Aquifer storage was developed in the Midwest to serve the large Greater Chicago market; deeper depleted gas fields were developed in Pennsylvania, Ohio, and West Virginia and the first bedded salt storage cavern was developed in Michigan. The first storage cavern in a salt dome opened in Mississippi in 1970 as a back-up supply needed after hurricanes. Today, the United States, Canada, and Europe house approximately 600 gas storage facilities (Figure 8.2).

Gas storage initially served as a single-turn baseload providing pipeline flexibility and users were charged for the service. The passage of the 1978 Natural Gas Policy Act, providing for escalating and, in some cases, incentive wellhead pricing, eventually allowed the industry to overcome a supply shortfall with very expensive deregulated intrastate gas. In recent years, interest in and demand for new storage have increased due to multiple factors, including the Federal Energy Regulatory Commission (FERC) Order 636 promulgating the end of most traditional pipeline merchant services. The result moved pipeline activity from suppliers to transporters, making

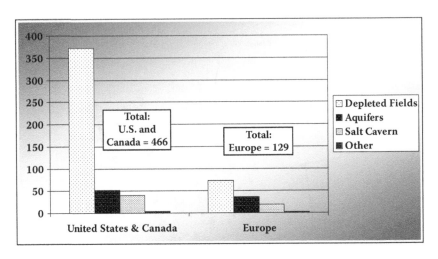

FIGURE 8.2
Underground storage fields in the United States, Canada, and Europe. (From American Gas Association and U.S. Department of Energy.)

long distance carriers (LDCs), electric generators, and large industrial users responsible for their own gas supply arrangements. Most storage owned and operated by the pipelines is redefined and unbundled from merchant services; thus, pipeline storage and transportation are no longer managed for end users, but operate for many diverse companies using common facilities.

The re-emergence of natural gas as an energy source of choice after the lifting of restrictions on its use as a fuel occurred primarily due to oversupply conditions arising from drilling in the early 1980s, the recession of 1983, the Clean Air Act movement, and increased imports from Canada. The result changed the profile of storage use from baseload 90 to 150 day deliveries and 200+ day re-injection periods to shorter durations and more flexible utilization.

Recent changes in the market for natural gas have changed the role and economics of natural gas storage. The development of market hubs increased the importance of strategically located storage fields and expanded the types of services available from storage providers. Changes in pipeline rate structures and the increase in importance of pipeline capacity discounting and capacity release altered the economics of many storage facilities. LDC restructuring is impacting the storage and contracting practices of many LDCs. In addition, the change in natural gas demand mix, particularly the growth in gas demand for electric generation, affected the overall load factor of natural gas demand, with a significant impact on storage requirements. These factors together have led to significant changes in natural gas storage operations.

Key Trends Influencing Future Value of Natural Gas Storage

Three significant market trends are changing the nature of natural gas storage markets. First, pipeline restructuring is changing the role and value of storage. Pipeline restructuring following FERC Order 636 led to market changes influencing the use and value of natural gas storage. These changes include the unbundling of pipeline service, a shift to straight fixed variable rate design, the development of a secondary market for storage and pipeline capacity, the growth in importance of market hubs and gas marketers, and a shift toward market-based rates for storage service (Beckman et al., 1995).

Second, the growth in gas usage for power generation is likely to increase the value of market area storage. Due to improvements in technology, favorable economics, and low emissions, natural gas is expected to be the incremental fuel for power generation for the foreseeable future. In addition, electric utility restructuring is resulting in the development of a flexible and visible spot market for electric power, greater incentives to minimize costs and increase operational flexibility of generating stations, increased flexibility, and incentives to arbitrage between natural gas and alternative fuels to generate electricity. These factors are likely to increase the value of high deliverability market area storage.

Finally, LDC restructuring is likely to increase the efficiency of storage services.

Natural gas LDCs are restructuring and unbundling in many states. The gas merchant function would be subjected to greater competition with services provided by gas marketers including unregulated affiliates of existing LDCs. At least initially, many more entities will hold the rights to pipeline and LDC storage. The holders of storage are more likely to have direct profit motives to maximize the value of the storage; hence, the market is likely to see new storage services offered and more effective use of existing storage facilities to increase profitability.

Types of Natural Gas Storage

The three major types of "reservoirs" common to the underground storage of natural gas are (1) depleted reservoirs, (2) aquifers, and (3) salt caverns. The East Coast region is characterized by depleted reservoir and aquifer storage; the Gulf Coast has a mix of depleted reservoir and salt cavern storage; the West Coast mainly uses depleted reservoirs. The components of a gas storage field are similar for depleted reservoir and aquifer facilities. Figure 8.3 illustrates the basic components (University of Michigan, 1978).

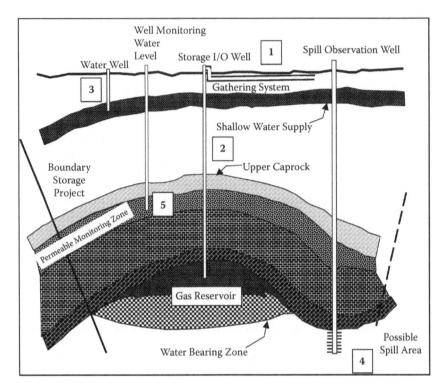

FIGURE 8.3

Gas storage system areas for natural gas monitoring. 1. Ground surface. 2. Wellbores. 3. Shallow water wells. 4. Reservoir spill point. 5. Permeable zone above caprock. (Modified from Katz, D. L., and K. H. Coats. 1968. *Underground Storage of Fluids*, Ulrich's Books.)

Depleted Reservoir Storage

The most common type of underground gas storage occurs in shallow, high deliverability depleted oil and gas reservoirs. Although the requirements vary, typically these reservoirs require 50% base gas (i.e., equal amounts of base and working gases) and present several advantages:

- They are near existing regional pipeline infrastructures.
- They have a number of usable wells and field gathering facilities to reduce the cost of conversion to gas storage.
- Their geology is well known; they contain previously trapped hydro-carbons that minimize the risks of reservoir "leaks."

Some disadvantages are associated with depleted reservoirs. Because of the nature of the reservoir-producing mechanisms, working gas volumes are usually cycled only once per season (extremely high deliverability storage reservoirs are exceptions). Often, these reservoirs are old and require substantial well maintenance and monitoring to prevent working gas from being lost via wellbore leaks into other permeable reservoirs.

Aquifer Storage

Aquifer storage involves injecting natural gas into underground formations that are initially filled with water (aquifers). The gas is injected at the top of the water formation and displaces the water down-structure. These types of reservoirs account for only 10 to 15% of total U.S. storage deliverability and exist mainly in the Midwest due to the lack of depleted oil and gas reservoirs. Advantages of aquifer reservoirs include:

- Proximity to end users
- High deliverability from a combination of high quality reservoirs plus water drive during the withdrawal cycle; high deliverability increases the ability to cycle the working gas volumes more than once per season.

One disadvantage is a high level of geological risk. These reservoirs have not previously trapped hydrocarbons and, as a result, a degree of uncertainty surrounds their ability to contain injected base and working gases. The risk for substantial reservoir leaks is also present. Because these reservoirs produce via water drive, water production is often experienced during the withdrawal cycle, increasing operating costs. Due to the water drive mechanism during the withdrawal cycle, the base gas requirements are high (80%). A large percentage of base gas is not recoverable after site abandonment. The high base gas requirement likely limits the number of new aquifer storage projects by increasing the initial capital cost (Katz, 1977).

Salt Cavern Storage

Salt cavern storage sites are solution-mined cavities in existing salt domes and structures (Figure 8.4). These shallow cavities are filled with injected natural gas and act as high pressure storage vessels—in essence they are large underground storage tanks for natural gas. Advantages include:

- Low base gas requirement of 25% that can approach 0% in emergencies.
- Ultra-high deliverability (much higher than depleted reservoir and aquifer storage).
- Operational flexibility: these reservoirs can cycle working gas four to five times a year. Their Gulf Coast location allows daily production and nightly injection to help meet peaking demands during the summer air conditioning season.

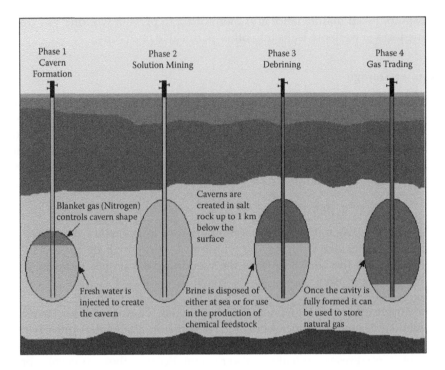

FIGURE 8.4
Salt bed gas storage cavern.

- Excellent seals: the cavern walls are essentially impermeable barriers.
- Small risk of reservoir gas leaks.

Disadvantages include:

- Long distance from the winter heating market to the north.
- Costly initial start-up: disposal of saturated salt water generated during solution mining may be costly and environmentally problematic.

The other types of storage contribute significantly smaller volumes of gas to the overall storage portfolio.

Liquid Natural Gas

Liquid natural gas (LNG) facilities provide delivery capacity during peak periods when market demand exceeds pipeline deliverability (Figure 8.5). LNG storage tanks possess a number of advantages over underground storage. Natural gas as a liquid at approximately −163°C (−260°F) occupies about

FIGURE 8.5
Gas storage field surface facilities.

1/600 of the space of underground storage and allows high deliverability on very short notice because LNG storage facilities are generally located close to market and the gas can be trucked to consumers, thus avoiding pipeline tolls.

Pipeline Capacity

Gas can be temporarily stored in a pipeline system via a process called line packing—adding more gas into a pipeline by increasing the pressure. During periods of high demand, greater quantities of gas may be withdrawn from a pipeline in the market area than are injected at the production area. Line packing is usually performed during off-peak times to meet the next day's peaking demand. This method, however, is only a temporary, short-term substitute for traditional underground storage.

Gas Holders

Gas can be stored above ground in a holder, largely for balancing purposes, not for long-term storage. Holders store gas at district pressure so they can provide extra gas very quickly at peak times. Gas holders are perhaps most common in the United Kingdom and Germany and are of two types. One type includes column guides that are always visible regardless of the position of the holder frame. Spiral guides have no frames and utilize concentric runners.

Role of Gas Storage in Transmission and Distribution

Underground storage is an essential component of an efficient and reliable interstate natural gas transmission and distribution network. The size and profile of the transmission system often depend in part on the availability of storage. Access to underground natural gas storage facilities, particularly those located in consuming areas, permits a mainline transmission pipeline operator to design the portion of its system located upstream of storage facilities to accommodate the level of total shipper firm (reserved) capacity commitments and the pipeline's potential storage injection needs (baseload requirements).

The segment of the transmission system downstream of the storage area (including LNG peaking facilities) is designed to accommodate the maximum peak period requirements of shippers, local distribution companies, and consumers in the area. It is generally sized to reflect the total peak-day withdrawal (deliverability) levels of all storage facilities linked to the system and estimated potential peak period demand requirements.

Some underground storage facilities are located in production areas at the beginning of the pipeline corridor and, in contrast to storage near consuming markets, can store gas that may not be marketable at the time of production. For instance, natural gas produced in association with oil production is a function of oil market decisions that may not coincide with natural gas demand or available pipeline capacity to transport the gas to end-use markets. Another example is the storage of natural gas produced from low-pressure wells that may be injected into storage during the off-peak season and delivered at high pressure to the mainline in the peak season.

In August 1991, FERC's Office of Economic Policy (OEP) released a discussion paper recommending that the commission improve natural gas market efficiency by recognizing and encouraging development of natural gas market centers (hubs), defined by FERC as places on the natural gas pipeline grid where buyers and sellers can make or take delivery of natural gas. FERC's definition requires that hubs must be near the intersections of several pipelines and, for convenience and balancing, should also be near fairly large production or storage areas. Order 636 promotes the use of hubs by requiring equal access to free markets for all gas industry participants through unbundling pipeline services and prohibiting pipeline tariffs from inhibiting the use of hubs (Beckman et al., 1995).

Two-thirds of the lower 48 states are almost totally dependent on the interstate pipeline system for their supplies of natural gas (Figure 8.6). On the interstate pipeline grid, the long-distance, wide-diameter (20 to 42 inch), high capacity trunk lines carry most of the natural gas transported throughout the nation. In 2007, more than 36 trillion cubic feet (Tcf) of natural gas were transported by interstate pipeline companies on behalf of shippers. The 30 largest interstate pipeline systems transported about 81% (29.8 Tcf) of the total. Natural gas is commonly routed through several interstate pipeline

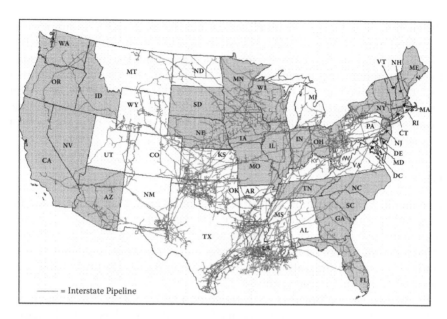

FIGURE 8.6
U.S. interstate pipeline system. Gray shaded states require pipeline systems for 85% or more of their total gas demands.

systems before it reaches its final destination. The interstate portion of the national natural gas pipeline network represents about 71% of the natural gas mainline transmission mileage in the United States. The 30 largest interstate pipeline companies own about 77% of all interstate natural gas pipeline mileage and about 72% (183 billion cubic feet [Bcf]) of the total capacity available within the interstate network.

Some of the largest pipeline capacities exist on natural gas pipeline systems that link the natural gas production areas of the Southwest with the other regions of the country. Sixteen of the 30 largest U.S. natural gas pipeline systems originate in the Southwest and 4 others depend heavily on supplies from the Southwest.

Customer Segments

Underground natural gas storage has traditionally been developed and used by pipelines and LDCs to optimize long haul capacity. Market area storage provided incremental city gate deliveries at the time required and in excess of long haul pipeline capacity. The natural gas industry invested in storage assets periodically as new market area deliverability was required. Prior to the enactment of Order 636, pipelines owned 59% of the nation's storage and managed most of their storage for system supply support. LDCs owned 38% of the national storage capacity and leased additional volumes from pipelines. The remaining storage was held by special purpose companies that generally offered sole use facilities to clients based on long-term leases. After Order 636 appeared, LDCs exchanged bundled gas services for comparable storage capacity. As a result, LDCs now control more than 70% of the current U.S. storage capacity.

LDCs

In general, LDCs continue to utilize most of their market area storage for seasonal baseload and peaking service. The key objectives for LDCs in evaluating the use of storage are supply security, peak-day coverage, and cost minimization. Ownership of the storage gas, proximity to points of consumption, and minimization of pipeline demand charges impact the LDC storage utilization decision process.

Production area storage is used by some LDCs as a substitute for more expensive swing supply arrangements. Gas is rarely held as a supply aggregation tool or for price arbitrage purposes. Production area storage is generally the first assigned to suppliers when LDCs trade capacity management for supply price discounts from long-term gas suppliers. High deliverability–high injection salt cavern projects have caught the imagination of all industry members except LDCs. Most such products want 75% or more financing of the $40 to $120 million price tags based on the credit qualities of the term lease storage clients the developers are seeking.

New production area reservoir storage projects are developing horizontal drilling technology as a method of enhancing reservoir injection and withdrawal characteristics to mimic salt cavern storage service offerings. This trend among reservoir storage developers reflects the perceived value of deliverability and flexibility of field area storage.

Suppliers and Aggregators

Gas marketing entities range from major corporations that take title to the gas to individual brokers living off buying and selling relationships. The larger supplier and aggregators offer extremely competitive swing gas supplies to LDCs, electric generators, and industries, often simply by manipulating daily gas flows and attempting to operate within the balancing practices of the pipelines.

Major marketers are aggressively entering the storage market as developers, clients, and agents. They are willing to invest in storage for the duration required to finance projects due to their need to establish the security of their supplies. They have hard assets in their portfolios in lieu of equity gas production and take advantage of the LDCs' willingness to pay swing charges rather than invest in storage. A storage position replaces wellhead gas supplies on the list of credentials that nonequity gas suppliers need to compete for value-added LDC business. The use of storage to support arbitrage and hedging of natural gas contracts is potentially the most profitable practice but also presents the greatest degree of risk. As a result, while the opportunity is available to LDCs, most do not pursue it because of restrictions of regulators.

Intrastate Pipelines

Intrastate pipelines have emerged as the most aggressive targets in the pipeline asset resale market. The value of storage for intrastate pipelines, whether procured as an equity investment or as a service lease, is essentially identical to the values set by large suppliers and aggregators. Value is defined by the contribution that the storage service brings to the gas sales function. This includes the value available to suppliers from swing supply sales, emergency back-up sales, balancing, no-notice service sales, and incidental income from arbitrage and peaking sales.

Interstate Pipelines

Storage has been a significant organizational function of interstate pipeline companies for many years. Open access and the elimination of merchant gas sales and services caused pipeline storage areas to effectively become "warehouse" operations. Order 636 allowed the pipelines to retain only such storage as required for operational integrity. Several major pipeline systems in the United States operate without the benefit of storage services for shippers or gas controls. These pipelines sell line packing mostly as a "no-notice" service, and they all attempt to cope with daily operating requirements through a variety of operational flow orders (OFOs) that give them sufficient latitude to require that shippers alter whatever behavior caused a problem such as insufficient gas entering the line to support deliveries or insufficient demand to match receipts.

Producers

In the past, producers acted as limited participants or nonparticipants in gas storage and marketing, primarily because of the corporate culture that pervades most major players that categorize their primary business as exploration and production (E&P). The E&P cultural issue is compounded by the fact that storage operators rarely earn more than a 15% return on after-tax equity employed, and E&P firms believe that their projects should have much higher hurdle rates to accept the high failure rate encountered in the drilling business. A few producers are examining storage as an adjunct to developing affiliate marketing organizations. Producers are now beginning to see some value in balancing, emergency back-up for warranty gas sales, and replacement sales volumes when platform production is interrupted by hurricanes or other factors.

Customer Segments Summary

Order 636 caused little change in the gas business. The same producers are selling gas in the market. The same pipelines are transporting it to the same end users. Some issues have changed. Who controls the pipeline capacity? How is it billed? Who is selling the value-added services available from storage (or comparable paper constructs sometimes called "virtual storage")? What are the rules for trading gas, pipeline, and storage capacity? Most importantly, who is ultimately responsible for gas service at the residential level? State level unbundling is challenging these factors and may constitute a far more compelling issue for the evolution of the storage industry over the next 10 years than federal regulatory change. The likely impact will be to make LDCs more peak-sensitive, require balancing across more shippers, result in continued disaggregation of utility loads, and shift gas supply responsibilities back to suppliers. In the meantime, the uncertainty of outcomes will likely preclude significant investment by LDCs in storage assets and may result in the growth of large suppliers and aggregators.

Economics of Storage

Underground natural gas storage presents a variety of economic justifications, depending on the perspective of the entity attempting to value a facility or service. In traditional rate making, the value attached to natural gas storage service was determined by the avoided cost economics of alternatives such as no-notice service. While avoided cost still plays a role for utilities, the value of natural gas storage is now evolving to valuing the services realized from expanded gas sales.

In general, as we see in the Figure 8.7 graph, high gas prices are typically associated with low storage periods. Usually when prices are high early in the refill season (April to October), many storage users adopt a wait-and-see attitude and limit their intake in anticipation that the prices will drop before the heating season (November to March). However, when that decrease does not occur, they are forced to buy natural gas at high prices. This is particularly true for local distribution and other operators that rely on storage to meet seasonal demands of their customers. On the other hand, other users

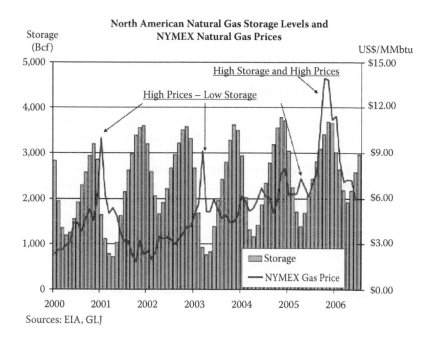

FIGURE 8.7

North America natural gas storage levels, 2000 through 2006.

who use storage as a marketing tool (hedging or speculating) will delay storing large quantities when prices are high.

As with all infrastructural investments in the energy sector, developing storage facilities is capital-intensive. Investors usually use the return on investment (ROI) as a financial measure to determine the viability of such projects. It has been estimated that investors require a rate of return between 12 and 15% for regulated projects and close to 20% for unregulated projects. The higher expected return from unregulated projects is based on higher perceived market risk. In addition, significant expenses are accumulated during the planning and location of potential storage sites to determine their suitability, and this further increases the risk. The capital expenditure to build a facility depends primarily on the physical characteristics of the reservoir. The development cost of a storage facility largely depends on the type of storage field.

As a general rule, salt caverns are the most expensive to develop on a billion-cubic-feet-of-working-gas-capacity basis. However, keep in mind that the gas in such facilities can be cycled repeatedly on a deliverability basis and thus may be less costly. A salt cavern facility may cost $10 million to $25 million per billion cubic feet of working gas capacity [4]. The wide price range is the result of region differences that dictate the geological requirements. These factors include the amount of compressive horsepower required, the type of surface, and the quality of the geologic structure.

A depleted reservoir costs $5 million to $6 million per billion cubic feet of working gas capacity. Another major cost incurred for a new storage facility is base gas. The amount of base gas in a reservoir may be as high as 80% for aquifers, making them very unattractive for development when gas prices are high. Salt caverns require the least base gas. The high cost of base gas is what drives the expansion of current sites instead of development of new ones. Expansions do not require much additional base gas. The expected cash flows from such projects depend on a number of factors such as the services the facility will provide and the regulatory regime under which it will operate. Facilities that operate primarily to take advantage of commodity arbitrage opportunities are expected to have different cash flow benefits from facilities used primarily to ensure seasonal supply reliability. Rules set by regulators can restrict the profits made by storage facility owners or conversely guarantee profits, depending on the market model.

Evolution of Storage

Storage has been and continues to be an important element in the U.S. natural gas supply portfolio. In the future, the value of storage will be driven by new market forces and paradigm shifts in the magnitude of short-term and seasonal gas price volatilities. The relative values of storage components will change. Seasonal storage capacity will become more valuable (wider magnitude of time spreads). Injection capacity also will become more valuable (reflecting need to balance large LNG vaporization send-out against variable and seasonal demand). Short-term deliverability may become less valuable because of aggregate increases in short-term and peaking deliverability from new LNG storage/vaporization facilities. Gas storage will be utilized to maximize unconventional gas recovery—gas production does not see the gas market. Gas-fired peaking power generation requires instantaneous short term gas supplies. Gas storage can be utilized to balance loads for variable renewable energy projects. Production outages will be mitigated via offshore gas supplies, large gas plants, pipelines, and other advances. As much as 650 Bcf of new working gas capacity may be required by 2020 (see Table 8.1) (Energy and Environmental Analysis, 2000).

Gas Storage Technology Development

Gas storage technology development has been conducted continuously since the evolution of the first storage fields. Better understanding of reservoir

TABLE 8.1

New North American Gas Storage Requirements

Incremental Working Gas Capacity	2004 to 2008 (Bcf)	2009 to 2020 (Bcf)	Total (Bcf)
Western Canada	30	40	70
Eastern Canada and Michigan	36	74	110
Midwest	–	60	60
New York	10	56	66
Pennsylvania and West Virginia	33	90	123
Gulf Coast	72	5	77
West Coast	21	78	99
Other	10	37	47
Total	212	439	651

Source: Base case from *At the Crossroads: Crisis or Opportunity for Natural Gas?* Energy and Environmental Analysis Inc. With permission.

responses to gas storage conditions, mechanical issues with wellbores, flow capacity, and formation damage have all been investigated. Gas storage presents some unique technology challenges as compared to other exploration and production operations. Included is the need to recharge storage fields on a regular basis. Recharging exposes the wellhead and wellbore environment to maximum operation pressures annually but requires all equipment to be maintained for those conditions. In standard oil and gas operations, the maximum pressure is realized during early production and declines thereafter.

Another area of unique challenge arises from aquifer gas storage fields that require overpressure conditions to inject gas into the reservoir. The operation needs a caprock to prevent gas movement from the reservoir during overpressure conditions. Typical oil and gas fields have proven the value of the caprock technique by the oil and gas accumulations contained. This condition does not exist for aquifer fields.

The U.S. Department of Energy conducted an industry workshop to assess technology needs for near- and long-term gas storage. Five major topic areas were identified and research needs within each category identified and prioritized (Table 8.2).

Gas Storage and CO_2 Sequestration

A number of technologies developed by the gas storage industry in the United States and Europe have potential application to CO_2 sequestration. The most utilized method of storing natural gas in geologic formations is injection into

TABLE 8.2

Gas Storage Field Technology: Near-Term, High Priority Research Needs

Reservoir	Mechanical	Water Issues	Data Management	Formation Damage
Develop methods to increase injectivity to provide increased cycling capability and/or reduced fuel usage.	Develop better understanding of maximum delta temperature that casings can withstand without failure of cement or joints.	Find new approach to handling produced water.	Develop low cost, low maintenance ±10% multiphase wellhead gas measurement system.	Continue investment in skin damage remediation technology. Remove N_2/CO_2 near well and wellbore for scales, fines, salts, asphalt, etc.
Develop technology to maintain existing path from formation to wellbore for gas flow.	Improve corrosion management methods to enhance availability (especially bacterial control). Develop new tools (e.g., logging) and techniques to verify integrity of casing strings.	Develop cost-effective means to remove water at end of withdrawal season. Develop ways to delay or prevent "watering off." Prevent water from encroaching on wellbore and reducing relative permeability.	Develop web-based data management tool to automatically store, archive, retrieve, and analyze routinely collected surveillance data.	Cost effectively identify and treat well damage mechanisms. Conduct basic research aimed at revealing most common reservoir damages.
Demonstrate by research that changing wettability increases gas deliverability. Develop innovative technologies (e.g., gas wettability) to increase capacity of existing storage fields at low cost.	Research improvements in deliverability by mechanical means such as new coil tubing tools.		Develop cross technologies and data mining for E&P.	Improve clean-out and simulation techniques to remediate damage.

Lost gas: condensates, migration, fractured reservoirs.
Develop technology to improve ability to interconnect formations via field experiments to demonstrate deliverability and interpret fuzzy logic.

Research gas to electricity concepts at peak production: boreholes, downhole fuel cells.

Optimize software that ties industry well data, hydraulics (pipe line simulation) characterization and system dispatch.

New approaches to modeling gas cycling to and from storage.

Improve lifetime and integrity predictions for gas storage wells to maintain capacity.

Correct wellbore damage.

Inventory verification: better techniques to handle changing uses of reservoir fields, e.g., average pressure.

Evaluate atypical compression and reservoir combinations for rapid in and out (4 ms) activity.
Ultrasonic meters.
Expand smart pipe concepts to production casing to prove feasibility

Determine near wellbore hydraulics.
Liquid banking.

depleted oil or gas reservoirs because these reservoirs have effective seals that prevented the escape of hydrocarbons for thousands of years. Thus the risks of losing stored natural gas are minimal (Benson et al., 2002).

However, depleted hydrocarbon fields in areas where natural gas storage fields are required are insufficient. The same is also is true for CO_2 sequestration where sites are needed in the industrial and highly populated areas, where depleted oil and gas fields are rare or nonexistent. The gas industry has overcome this obstacle in part by creating storage fields in aquifers, and this technique is an obvious choice for sequestration of carbon dioxide in many industrial and highly populated regions around the world.

Storage of natural gas in aquifers is the process of injecting gas into an aquifer of high or reasonably high permeability under structural conditions that mimic natural oil and gas reservoirs, for example, anticline highs or up-dip pinch-outs. In addition, a target aquifer must be free of faults so that the stored gas will not escape through fault planes.

The keys to the success of storing natural gas and/or carbon dioxide in geologic formations are site selection and accurate delineation of the host formation to ensure that the formations are continuous and extend over a wide area without faults or other discontinuities that would allow escape of the injected gas. A storage zone must be contained below impermeable overlying beds, preferably structurally undisturbed, and laterally continuous to store large quantities of gas to be injected continuously over a very long period. In addition, for any method of gas storage or carbon sequestration to have value, a reliable monitoring procedure must ensure that the process follows the projected path. Monitoring must also implement early remedial action when required.

A number of technologies developed by the gas storage industry in the United States and Europe have potential application to CO_2 sequestration. Table 8.3 identifies those technologies.

Successful CO_2 sequestration requires participation by many disciplines. The technology developed by the underground gas storage industry will have significant application to CO_2 sequestration. Gas storage operators have developed a technology portfolio that is not widely known or generally available across the E&P industry. The availability of this technology and increasing the awareness of these possibilities could prevent duplication of effort, resulting in considerable savings and providing more effective CO_2 sequestration (see Figure 8.8).

The gas storage industry has successfully operated storage fields for over 90 years and has developed a number of procedures and technologies that can directly assist with the sequestration of CO_2. Many of the technologies utilized by gas storage operators were developed by and adopted from the oil and gas industries.

Several technologies and procedures have been developed by the gas storage industry directly to meet customer needs. This is especially true in the area of aquifer gas storage which includes a portfolio of technologies unique to the aquifer storage business. All the existing gas storage technology can

TABLE 8.3

Gas Storage Technologies with Potential Application to CO_2 Sequestration

Technology Area	Gas in Place Determination	Leak Detection	Leak Control	Gas Movement Monitoring	Caprock Integrity Determination	Sealing Caprock Leaks	Reservoir Suitability For Storage
Pressure–volume techniques	x	x					
Reservoir simulation	x			x			x
Volumetric techniques	x	x					x
"Watching the barn doors"	x	x		x	x		
Surface observations		x	x	x			
Changes in vegetation		x	x	x			
Shallow water wells		x	x	x			
Observation wells	x	x	x	x	x		x
Well logging	x	x	x	x			x
Seismic monitoring		x		x		x	x
Gas metering	x			x			
Gas sampling and analysis		x		x			
Tracer surveys		x	x	x			
Production testing	x	x			x		
Remote sensing		x	x	x			
Shallow gas recycle			x				
Aquifer pressure control			x				
Caprock sealing techniques		x	x			x	
Geologic assessment	x	x		x	x	x	x
Threshold pressure					x		x

(Continued)

TABLE 8.3 (CONTINUED)

Gas Storage Technologies with Potential Application to CO_2 Sequestration

Technology Area	Gas in Place Determination	Leak Detection	Leak Control	Gas Movement Monitoring	Caprock Integrity Determination	Sealing Caprock Leaks	Reservoir Suitability For Storage
Pump testing		x			x		x
Flow and shut-in pressure tests		x			x		x
Air and CO_2 injection		x			x		x
Wellbore damage							
Overpressuring			x	x	x		x

Type	Development Costs per Bcf of Working Gas Capacity
2-Cycle Reservoir	$5 – $6 million
6-to-12 Cycle Salt Cavern	
Gulf Cost	$10 – $12 million
Northeast and West	As much as $25 million

FIGURE 8.8
Costs of development of working gas storage per billion cubic feet of working gas capacity. (From Federal Energy Regulatory Commission Report Docket AD04-11-000.)

provide significant insights for the CO_2 sequestration industry. The aquifer gas storage area may have the greatest contribution to make to the CO_2 sequestration business (see Table 8.3). Gas storage operators have accumulated a significant knowledge base for the safe and effective storage of natural gas. While unwanted gas migration has occurred because of mechanical problems with wells and geologic factors, gas storage overall has been effectively and efficiently performed.

References

Beckman, K.L., Determeyer, P.L., and Mowrey, E.H. June 1995. Natural Gas Storage: Historical Development and Expected Evolution. International Gas Consulting, Inc., Houston. GRI 95/0214.

Benson, S.M. et.al. 2002. Lessons Learned from Natural and Industrial Analogues for Storage of Carbon Dioxide in Deep Geological Formations. Carbon Capture Project.

Energy and Environmental Analysis, Inc. February 2000. Natural Gas Storage: Overview in a Changing Market Environment., Arlington, VA. GRI-99/0200.

Federal Energy Regulatory Commission. September 30, 2004. Staff Report. Docket No. AD04-11-000.

Katz, D.L. 1977. Making Good Use of Observation Wells in Gas Storage, American Gas Association Operating Section Proceedings.

Katz, D.L., and Coats, K.H. 1968. *Underground Storage of Fluids*. Ulrich's Books.

University of Michigan. May 1978. Proceedings of Symposium on Underground Storage of Gases, Engineering Summer Conference, Ann Arbor.

Additional Sources

Burnett, P.G. 1967. Calculation of the leak location in an aquifer gas storage field. Society of Petroleum Engineers Gas Technology Symposium, Omaha.

Katz, D.L., 1971. *Monitoring Gas Storage Reservoirs*, Society of Petroleum Engineers Preprint 3287.

Katz, D.L. 1978. Containment of Gas in Storage, American Gas Association Operating Section Proceedings.

Katz, D.L., Elenbaas, J.R., and MacDonald, R.C. June 1983. Natural Gas Engineering Production and Storage.

Katz, D.L. et al. 1959. *Handbook of Natural Gas Engineering*. McGraw Hill, New York.

Thomas, L.K. 1968. Threshold pressure phenomena in porous media. *Society of Petroleum Engineers Journal.*

Witherspoon, P.A. July 1967. Evaluating a slightly permeable caprock in aquifer gas storage: caprock of infinite thickness. *Journal of Petroleum Technology.*

Witherspoon, P.A., Javandel, I., Neuman, S.P. et al. 1967. *Interpretation of Aquifer Gas Storage Conditions from Water Pumping Tests*, American Gas Association, New York.

Witherspoon, P.A., Mueller, T.D., and Donovan, R.W. May 1962. Evaluation of underground gas storage conditions in aquifers through investigation of groundwater hydrology. *Journal of Petroleum Technology.*

http://www.neo.ne.gov/statshtml/124.htm

Index